The Local Government Guide to Imaging Systems: Planning and Implementation

A publication of Public Technology, Inc., and the International City/County Management Association

In cooperation with

Atlanta, Georgia
Austin, Texas
Hamilton County/Cincinnati, Ohio
Hennepin County, Minnesota
Houston, Texas
Memphis, Tennessee
New York, New York
Palo Alto, California
San Diego, California
Santa Rosa, California
Scottsdale, Arizona

Research supported by

Andersen Consulting
Booz-Allen & Hamilton
Eastman Kodak Company
PRC, Inc.
Unisys Corporation
US West

Public Technology, Inc. (PTI), is the non-profit technology research, development, and commercialization arm of the National League of Cities, the National Association of Counties, and the International City/County Management Association. Through collective research by its members, PTI creates and advances technology-based products, services, and enterprises in cities and counties nationwide.

PTI's membership of 150 innovative local governments includes the Urban Consortium (UC), a network of jurisdictions with populations of over 250,000. The UC provides a platform for research and enterprise through its Energy, Environmental, Transportation, and Telecommunications and Information Task Forces.

ICMA is the professional association of appointed administrators serving in cities, counties, regional councils, and other local governments. Its primary goals include strengthening the quality of local government through professional management and sharing information on new practices and approaches to governance through a wide range of information services, publications, and training programs.

Through its comprehensive research and analysis activities, ICMA makes available newsletters, reports, books, and specialized research-assistance subscription services that enable local government officials to learn from the experience of other practitioners.

ISBN: 0-87326-097-X
Printed in the United States of America.
9 9 9 8 9 7 9 6
5 4 3 2

Table of Contents

Acknowledgments

In 1993, Public Technology, Inc. (PTI), launched a unique, joint public-private research project called Local Government Imaging Applications and Strategies. The goal was to bring experienced cities and counties together with expert private-sector firms to identify, examine and discuss the policy, management and technology issues that relate to imaging systems. This guidebook is the result of those efforts.

Twelve jurisdictions and six private-sector firms sponsored and participated in the project. We wish to thank the following individuals for their contributions and for reviewing the draft of this book: From **Atlanta, Georgia**, Jim Bishop and Ellis J. Colbert; from **Austin, Texas**, Pat Miller and Dean LaBonte; from **Hamilton County and Cincinnati, Ohio**, Steve Schutz; from **Hennepin County, Minnesota**, Gary Yochum and Jackie Weiler; from **Houston, Texas**, Michael Antash and Louis Aulbach; from **Memphis, Tennessee**, Claudia Shumpert; from **New York, New York**, Al Leidner and Cecil McMaster; from **Palo Alto, California**, Dianah Neff; from **San Diego, California**, Jeanne Culkin and T.J. Murray; from **Santa Rosa, California**, Ron Kaetzel; and from **Scottsdale, Arizona**, Jeff Denning.

Special thanks go to the following representatives from the project's six private-sector sponsors: From **Andersen Consulting**, Greg Carney; from **Booz-Allen & Hamilton**, Reggie Lawson; from **Eastman Kodak Company**, Marilyn Sadler-Bay; from **PRC, Inc.**, Robert E. Marggraf; from **Unisys Corporation**, Bill Beckham, Jim Brown and Frank Iacono; and from **US West**, Michael Jordan and Marjorie Stoltz.

Inspiration and initial direction for this project came from members of PTI's Urban Consortium Telecommunications and Information Task Force, which identified imaging technology as a research priority for local governments, recruited private-sector sponsors for the R&D project and suggested project participants from PTI's membership of innovative cities and counties.

PTI's President, Dr. Costis Toregas, and several former and current PTI staff—Francie Gilman, Mike Humphrey, Cindy Kahan, Ted Shogry, Shaden Tageldin and Taly Walsh—provided research, editorial and production guidance.

About the Author

Tod Newcombe is a freelance writer specializing in information technology and telecommunications. Since 1990, he has written dozens of articles on technology for numerous publications, including *Government Technology*, *Governing*, *Imaging World*, *Inform*, *LAN Magazine*, *Network World* and *Urban Land*. Prior to working as a freelance writer, he was communications manager for PTI. He lives in Longmeadow, Massachusetts.

Chapter 1

Introduction

Despite the growing quantity of data that exists today in electronic form, paper remains the dominant medium by which local governments store and process information. In fact, governments and businesses store a very small percentage of their information on computers, microfilm or magnetic tapes, according to the Association for Information and Image Management (AIIM).

Those figures are beginning to change—and with good reason. Imaging technology is making its way into every level of local government, converting paper documents into electronic images that computers can store, retrieve, display and disseminate far faster and better than any human being, changing forever the way work is processed. The change may unleash significant benefits for cities and counties in productivity and cost savings. More important, imaging can invigorate services by making information accessible to citizens as never before.

The importance of imaging to local governments cannot be overstated. Cities and counties still maintain 90 percent of their information in paper documents, according to some estimates. Much of this information originates outside of local government, making it hard to control how the information is captured and formatted. Think of the land records and titles that individuals and lawyers file every day. Or the birth and death certificates submitted by hospitals and doctors, and the numerous tax filings from individuals and businesses. Then the building permits from contractors and licenses from hunters that pour in daily. The list goes on. Little, if any, of the information is submitted electronically to local governments. What information is submitted on paper can be handwritten or typewritten in dozens of different ways.

With no controls, local governments have few choices about how to manage their documents efficiently. As a result, cities and counties must budget precious tax dollars for space and labor to manually store, retrieve and process the paper documents. In fact, a large percentage of government resources are expended on tasks, such as filing and storing, that add no value whatsoever to the information on the paper.

With imaging, however, cities and counties have a tool that can tame the paper beast, allowing departments to devote more time to urgent problems and derive more value from the information contained in documents. The use of imaging in government grows every day. With proper planning, departments are succeeding in turning back the tide of paper, slashing costs, boosting efficiency and even earning revenue.

Numbers and Pressures

In recent years, the quantity and cost of paper documents produced, received and managed by local government have risen considerably.

Consider these facts:

- Each working day it is estimated that more than a billion paper documents are generated in the United States alone.[1]
- Some three percent of this total is misfiled,[2] costing an average of $140–$200 per document to recover.
- The number of paper documents produced annually in the U.S. is expected to reach 1.6 trillion by 1996.[3]
- The cost of floor space has gone up 300 percent in the last decade.[4]

For city and county governments, with cash-strapped budgets, the idea of committing an ever-increasing amount of resources to store and manage paper documents is becoming unacceptable.

Another concern with paper-based documents is their impact on public service. Retrieving paper documents is a slow process. In the past, waiting hours, days or even weeks to receive a copy of a public document used to be the accepted norm in doing business with government. But today, that's no longer the case. Taxpayers, as customers, have come to expect rapid service response. Because of automation and other improvements in the private sector, including toll-free service numbers, fax and overnight delivery, individuals are no longer willing to wait very long for a service to be rendered. As a result, local governments are under increasing pressure from businesses and citizens to become more responsive.

Key hardware components of an imaging system include the server, which handles software and databases; the desktop computer, for viewing the images and performing computing tasks; and the optical jukebox, for storing and accessing the optical disk platters that contain the document images.

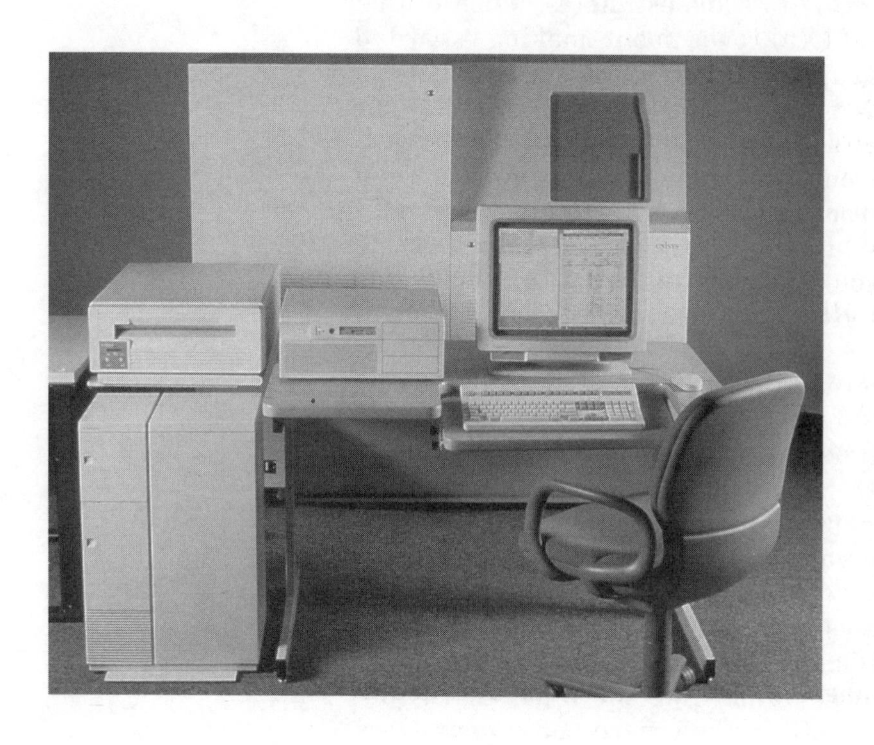

Imaging Definition

The technology for managing local government documents has, until recently, only inched forward. For many years, leather-bound volumes and filing cabinets were the standard tools for document storage. Micrographics, which has been around for decades as microfilm and microfiche, expanded the quantity of paper documents a local government could store and lowered storage costs. Micrographics is still used extensively in government today. (See "Micrographics," Chapter 4.)

In the 1980s, computers and scanners made it possible to create electronic images of pictures and paper. A scanner is a device that "takes a picture" of a document, creating a TV-like matrix of dots called a bitmap. The bitmapped image is quite large when compared to a word processing or spreadsheet file. The image can be stored on magnetic disk in a computer or on an optical disk, using laser technology.

Magnetic disks are the primary means of storing data on computers today. They provide fast access and can both store and erase files, but they are a relatively expensive method for managing large quantities of data. Optical disks provide a slower but less expensive option for storing data such as document images.

One optical disk can hold 40,000 to 100,000 pages of documents. When many disks are mounted in an optical jukebox—a robotic device roughly the size of a refrigerator that can mount and dismount as many as 100 to 200 disks—they can store the same amount of information as hundreds of four-drawer file cabinets. Because of technology's constant improvements, the number of pages an optical disk can support and the number of disks a jukebox can mount will continue to increase.

To the computer, the scanned image of a paper document is a graphic file, which has meaning only to people when they see it on a computer screen. If people are to search for and use the contents of the document image on a computer, the image must either be converted into computer-readable data or indexed for easy retrieval. A technology called optical character recognition (OCR) can convert images of letters and numbers into code the computer can read and understand.

However, OCR capabilities are limited by the quality of the character images and whether they are typewritten or handwritten. Given these restrictions and the fact that many documents stored by government—such as deeds or birth certificates—must retain their original format, most scanned documents are saved as images, checked for quality and indexed by a computer operator using key fields before they are stored. The index allows the user to find the document image quickly.

Software links the individual hardware components—scanner, computer, jukebox and printer—and enables users to search, retrieve, view, distribute and print the document image. Today's imaging software is designed to operate across a network of computers, so that offices, departments and even entire organizations, or enterprises, can share and access document images.

The software also enables the images to be integrated into an existing computer system. This means, for example, that a mainframe computer with an existing database of birth and death information can be linked with images of the actual birth and death certificates. Should an individual request a copy of a birth certificate, a government worker merely has to search the database for the person's name and, when found, have the software retrieve the certificate image and print out a copy.

Technical Issues

Information entered into a computer using a word processing, database or spreadsheet program can be edited, analyzed and tracked because each character is represented by a computer code called ASCII (American Standard Code for Information Interchange) that responds to commands from the software programs. A document image, however, is merely a series of light and dark pixels, which form a bit-mapped image file. If the image is of a page of text, the computer has no control over the characters that appear in the image.

A one-page document in ASCII code takes up about 3,000 bytes or three kilobytes (3K) of storage in a computer. An uncompressed scanned image of the same one-page document can occupy as much as 500,000 bytes or 500K of storage. Even when compression technology is used, the image still occupies a much larger amount of space in a computer's storage system than character-coded text.

Open Systems

The movement away from proprietary computer systems has progressed over the past several years in response to customer demands for interchangeable hardware, software and data. This ongoing trend, best exemplified by the large number of choices in personal computers and software programs, is called open systems.

In open systems, different brands and models of mainframes, minicomputers and personal computers and their software can share information. This shift away from single-vendor systems gives government managers freedom to choose the most cost-effective computing strategy that fits their department's needs.

The key to open systems is standards, which provide customers with the latitude to select from a broad range of hardware and software that is interoperable. Open systems standards protect a customer's investment in software, hardware and training, while allowing him or her to add enhancements as required. Overall, open systems standards can reduce costs and provide greater access to government data, allowing individual departments to tailor computers to their specific needs, while maintaining data security.

Some examples of standards include the Intel microprocessor chip, the PC operating system and user interface MS-DOS and Windows, and the workstation operating system called UNIX. In imaging, the most significant standards to date are those for compression of black-and-white, grayscale and color images and for transferring images between different brands of imaging software.

Because of their size, document images can have a significant impact on a computer system's performance and equipment needs. For example, while everyone with a computer can view a text file on his or her monitor, not everyone can view a document image on his or her computer. In order to do so, the computer must be "image-enabled" so that it can display the image at the required high resolution.

Users must also consider the technological issues of sharing images. Today's computer networks are designed to transfer the relatively small amounts of data that make up word processing, electronic mail and spreadsheet files. Adding large imaging files to a network can slow down data transfer considerably. Existing networks can probably handle a small amount of document image-sharing, but if the imaging application is large or expansion plans are in the wings, the network will have to expand to handle the additional capacity.

Big vs. Small

When imaging first became available, the technology could only be cost-justified as a large system serving many users. Imaging systems were built to run on mainframes or minicomputers using proprietary hardware and software. Today, an imaging system can consist of just one scanner, one computer, one optical storage device, one printer and one user. Most imaging technology runs on industry-standard personal computers instead of on proprietary systems, and users have the freedom to mix and match the types of hardware and software they wish to use.

While the new open systems method of computing has made imaging available for applications of all sizes, some imaging programs are better suited for high-end, large-scale applications, while others handle smaller needs much better.

As expected, high-end imaging systems offer a much greater range of features and are better suited for complex applications, such as processing medical claims and tax forms, or managing land records and court cases in a busy court. Some of the features included in

An image server is the computing workhorse of an imaging system, providing the processing power for everything from scanning and indexing to document retrieval and database management. Large image servers are typically used in high-end imaging systems that have a large range of features and complex applications.

large-scale imaging systems are methods for speeding up the delivery of images from the optical disks to the user's computer, faster printing of images and other performance-enhancing capabilities, such as workflow software that facilitates the routing of electronic documents between personal computers. High-end systems also provide a greater degree of customization than low-end systems.

Low-end systems, which can cost several thousand dollars less per user than high-end systems, target simpler problems and smaller workgroups. Typical applications might range from a mayor's or county executive's office correspondence to police reports for a small law enforcement agency. While low-end systems boast fewer features than their high-end counterparts, more vendors sell low-end imaging systems.

Imaging at Work in Local Government

Imaging technology has been implemented in a variety of departments in city and county government. To date, however, the biggest practitioners of imaging have been those departments with the largest record files. Land records offices, vital records divisions, police departments and courts—among the earliest users of imaging—have the longest track record with the technology. Here are a few examples:

Land Records

Registry offices are ideal users of imaging technology. They receive large quantities of paper documents from the public that must be catalogued, filed and made available for examination. A growing number of land registry offices are implementing imaging to automate the storage and retrieval of documents, provide faster public access and allow outside access to the database of images as a value-added service.

Imaging Opportunities Exist When:

- Documents originate outside the organization
- Movement of paper is a bottleneck
- Access to paper is too time-consuming
- Retrieval delays limit opportunity
- Document processing is manually intensive
- Physical routing introduces processing delays
- Transactions take a long time
- Unavailable documents expose organization to risk
- Document integrity and security are important
- Simultaneous access is important
- Paper-based operation is critical to organization

Source: Unisys Corporation

In 1989 the Middlesex County, Mass., Registry of Deeds installed an imaging system to automate the storage and retrieval of mortgages, deeds and liens. With all documents filed since 1960 now scanned and saved on optical disks, the Registry can retrieve, display and print the majority of its active documents for homeowners, real estate brokers, bankers, attorneys, title examiners, assessing officers and historians in seconds instead of minutes or hours. The system is designed so that people with modems—devices that allow computers to communicate over telephone lines—can look up documents on the Registry's database from their offices and request the Registry to either mail or fax them a copy of the document.

Police Records

Increasing criminal activity, as well as state and federal mandates on crime reporting, has forced even the smallest law enforcement agency to divert scarce resources to records management. More problematic is the amount of time officers and investigators must spend waiting for files on active crime cases. With imaging, police departments are reporting the elimination of paper bottlenecks, reduced clerical needs and improved access to police report files.

The Stockton, Calif., Police Department used to rely on colored-paper photocopies and pneumatic tubes to distribute active crime reports to its various divisions. The Department estimated that 2,000 hours of clerical time were spent each year chasing down misfiled documents. In 1992, the department installed an imaging system with workflow capabilities. When scanned documents are indexed, clerks are prompted by the system to answer questions about the documents' subject matter. Based on the answers, the imaging system determines where electronic copies of the report will go, and automatically routes them to the appropriate divisions.

Courts and Parking Bureaus

Like land registry offices and law enforcement agencies, courts are heavily dependent on paper documents and can benefit from imaging technology. To cost-justify the large investment necessary for imaging, some court systems are first implementing the technology where the return is the greatest. Parking enforcement is one such area. Imaging can facilitate the collection of parking fines and offer violators better service if they choose to contest a ticket.

Enforcement of Chicago's parking ticket program used to be so lax that the city was able to collect on only 10 percent of its parking tickets. The city's failure to enforce its ticketing program also increased traffic and pollution in the city's celebrated downtown Loop district. To remedy the problem, the city took enforcement out of the central traffic court and dispersed it to neighborhood hearing centers. An imaging system was added to scan the more than 12,000 tickets issued daily, as well as correspondence from violators. Each hearing center has imaging workstations that hearing officers use to retrieve and display tickets either for processing or for a hearing when the violation is contested. Since the centers are connected via a computer network that shares the ticket images, violators can visit any center to resolve a problem.

Imaging Applications in Local Government: A Matrix

Application	Why Important	Lead Department	Customers	Potential Revenue	Difficulty	System Interfaces
Development Plan Process	Improved customer services; reduction of complexity	Planning; Building Inspection; Engineers	Anyone who wants to build	Good; permit fee increase	High	Minimum of three: planning, inspections, engineering
Real Estate	Efficiency	Assessor's Office	Code Administrator; Planning; Assessor; Public	Modest	Modest	Real estate and building permits
Crime Adjudication Documents	"80 percent of misdemeanors are freed due to 48-hour processing requirement."	Police; Attorneys; Judicial System	Citizens; Police; Court; Defendants; Victims	None	High	Criminal Justice Information System (CJIS)
Criminal Records	Public safety; productivity; customer service	Police	Police; State/Federal Governments; Public; Attorneys; Media; Insurance	Medium/Low	Easy/Medium	State/fed access; field officer; records management system; CAD
Accounts Payable Vouchers	Efficiency; retention requirements; cost avoidance	Controller	All Departments	None	Easy	Purchasing
Medical Records	Efficiency; health; reimbursement tracking	Health; Human Services; Hospital	Internal; Patients; Doctors; Lawyers; Insurance Companies	High	High	Billing; registration; order entry; patient care
Housing Rehabilitation Loans	Reduction of turnaround; cost avoidance	Housing; Community Development	Citizens; Housing Inspectors; Planners	None	Low	GIS; citizen complaints; code enforcement
Legal Case Management	Productivity; cost avoidance; increased fees; efficiency	Law	Attorneys; City Manager; Risk Managers; Administrative Law Proceedings	Low	Medium	Office systems; risk management; worker's compensation
Human Resources/ Personnel	Management of paper-intensive work; cost savings	Human Resources	Applicants; All Departments	None	Medium	Position center; finance; legal
County Recorder/ Public Records	Productivity; low risk	County Recorder	Title Companies; Attorneys; Examiners; Assessors; Real Estate; Auditors; Public	High	Easy	Mainframe index; GIS
Adult Field Services/ Parole	Productivity; public safety	Corrections Department	Corrections; Courts; Parole Officers	None	Medium	Client/server (adult services)
Construction Contract Tracking	Productivity; quality control; cost avoidance	Environmental Protection; Engineering; Public Works; Utilities	Agencies; Utilities; Construction Companies	None	Medium	WAN; remote with private sector; planning department
Utility Billing and Customer Correspondence	Customer service	Administrative Services	Citizens; Customer Service	None	Easy	Existing utility billing system
Tax and License Records	Information sharing; customer service	Financial Services	Police; Customer Service; Planning; Zoning; Tax Auditor	Low	Easy	GIS; existing systems

Source: Local Government Imaging Applications and Strategies Project

The Benefits of Imaging

The benefits of imaging in local government are as far-reaching as its many uses. Often cited first is the ability to save valuable office space and storage costs by converting large volumes of paper stored in file cabinets into electronic images permanently stored on optical disk platters. Many paper-swamped government agencies have sacrificed valuable working space to store documents. With imaging, they can win back the space to the benefit of their employees.

Imaging keeps documents in a secure environment where they can't be misfiled or stolen. The scanned image of a document can be permanently affixed to an optical disk for long-term storage, protected from possible tampering or accidental erasure.

Once a document is scanned, it is accessible to anyone on a network with a computer capable of displaying images. The access can be simultaneous, so that more than one person can view the document at the same time. Access time can be counted in tens of seconds, not tens of minutes, as is the case in a typical paper records management environment.

Simultaneous access of document images leads to what many consider the technology's most significant benefit—the ability to eliminate tasks that exist solely for processing paper. In any kind of paper-based environment, documents move serially from one worker's desk to another upon completion of the task. With imaging, workers no longer have to wait for someone else to complete a task before they start theirs. Document images move in parallel fashion, allowing for greater worker efficiency. Innovative users of imaging have re-worked or re-engineered tasks and staff duties and have cut costs while boosting productivity.

When the Stockton Police Department automated the flow of crime reports with imaging, it was able to absorb an increase in reports without increasing staff. The city of Chicago's decentralization of its parking ticket program was a direct result of imaging. Hearing officers no longer needed to station themselves next to files containing tickets and correspondence in a central, downtown office. Imaging enabled them to move out to where the violators lived—in the neighborhoods of Chicago. Imaging enabled the Syracuse, New York Police Department to cut clerical labor costs by $200,000 and put two officers back on the streets.

Imaging has also proven successful in bringing in badly needed revenue. Chicago's parking ticket revenue soared from $34 million in 1989 to $48.8 million in 1991 after imaging was introduced. Middlesex County's Registry of Deeds charges fees to businesses that look up documents on its database and have images faxed to their offices. The county has been averaging between $80,000 and $100,000 in revenue annually since its imaging system was installed.

Finally, imaging can benefit public service delivery. In any local government department where imaging is in place and a public service is provided, you will find a public-access terminal where an individual can locate a document in seconds instead of waiting to be served. These kinds of service enhancements can't be dismissed as frills. Imaging is part of the wave of electronic service delivery that's sweeping the public sector. With tighter budgets, cities and counties can no longer afford to add more front-desk clerks to boost service. When customers can help themselves at a computer capable of displaying document images, a clerk is free to handle more complicated tasks. When imaging can be used to provide a specialized service, such as Middlesex County's deed-on-demand service, customers are often willing to pay more for the benefit.

Success Factors

With so much to gain from imaging, it is easy to overlook its potential problems and pitfalls. Imaging remains a complex and costly technology that is still relatively new, especially in the area of standards. Gaining the maximum benefit from imaging calls for extensive planning at both the policy and management levels of government, as well as a firm understanding of the technological variables. Overlooking any one of the crucial planning steps can break budgets, slow down implementation and cut short anticipated benefits.

Certain key actions determine whether an imaging system will become successful:

- **Begin by setting strategic goals and objectives.** Effective implementation of imaging requires a common vision of what the city or county expects from the technology and a strategy for meeting those expectations. Strategic planning can be one of the most effective ways to define goals and set a path for reaching them. With strategic planning, a city or county can establish where imaging opportunities exist, make informed decisions about its goals and plan a realistic way to attain them.
- **Engage a top-level champion to ensure success.** Planning a successful imaging system requires strong management support. Imaging involves not just a new technology, but information and how it is processed, tasks and how they are performed and the people who perform them. A project leader can see that all these elements receive equal consideration. If the use of imaging is to be cross-departmental, a leader can settle any factional disputes that might arise over such matters as data ownership, hardware and software standards and other issues.
- **Find people with the proper skills.** Imaging presents unique technological and managerial challenges. Most government departments have limited experience with client/server computing, scanners, relational databases, OCR and other hardware and software components unique to imaging. Similarly, management tools, such as business process re-engineering and change management, are not widespread in government, but are important to the success of imaging. Be sure the imaging project involves people—inside and outside government—who have the skills to use these tools.

- **Consider usage factors, legality of images, and document retention requirements.** Documents that are rarely used probably don't require imaging. A lower-cost solution, such as microfilm, might be the answer. More important, review policy concerning retention limits and the legal ramifications of storing certain public records as electronic images rather than as paper documents. Finally, if converting existing paper files, examine the documents and ascertain which should be saved. The cost of converting existing files into images is significant, often exceeding the cost of the imaging system itself. Elimination of redundant or unimportant documents can lower conversion costs.
- **Be sure to start with a winning application.** According to most surveys and studies, imaging will become a predominant technology in local government in the coming years. However, the speed with which it spreads throughout a city or county will depend on the success of the first application. City or county councils will be reluctant to fund expansion of imaging if the initial project takes too long to implement, costs too much or falls short of its projected goals. Select the most appropriate application, one where the payback potential is the greatest. If the approach is to start with a pilot system, be sure it's part of the overall solution strategy, not just an attempt to understand the technology. Many pilot projects have died because they didn't fit in strategically.
- **Devote time and resources to change management and re-engineering.** Because so much information in local government resides on documents, changing from a paper-based information system to one that is automated by imaging can be challenging to an organization and disruptive to its workers. Communication patterns will change as routine tasks are automated. Some employees will use computers for the first time. In order for imaging to succeed, the organization must help its staff absorb these changes. With change management techniques and tools, government departments can address change successfully.

  Re-engineering is now considered by many to be an essential step in implementing any strategically important imaging system. It provides local governments with the greatest opportunity to significantly boost productivity, enhance services and cut costs. However, the process of re-engineering is not for the faint-hearted. It involves completely redesigning how work flows through a department or an entire government. Re-engineering also calls for the elimination of many traditional tasks and the implementation of entirely new ways of doing business. Handling the human factor in a re-engineering project can make choosing an imaging system look easy. Experts say that candidates for re-engineering need to start their work well in advance of implementation and must be willing to devote the resources needed to get the job done properly.

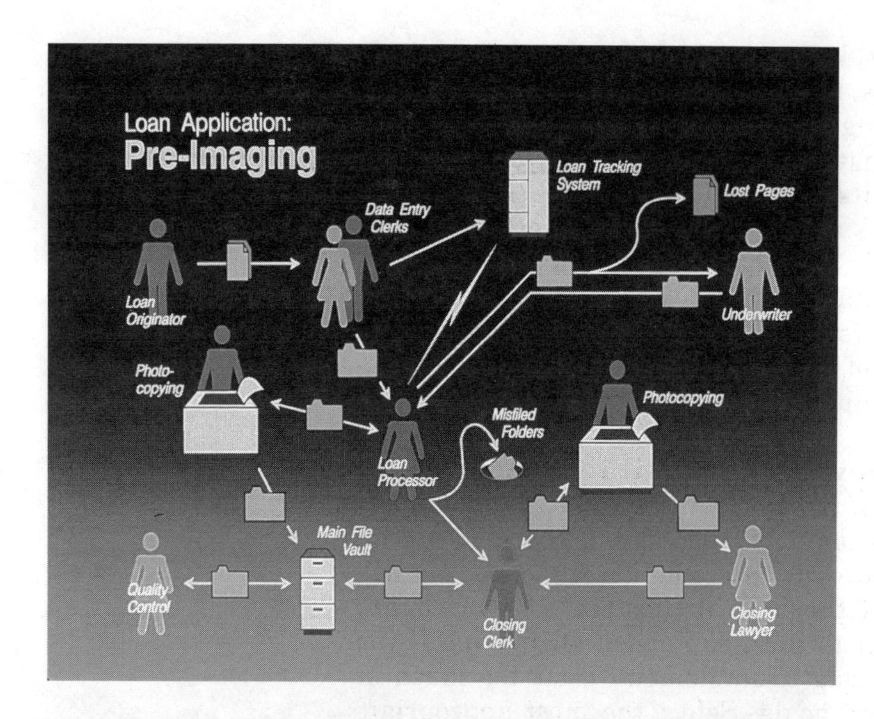

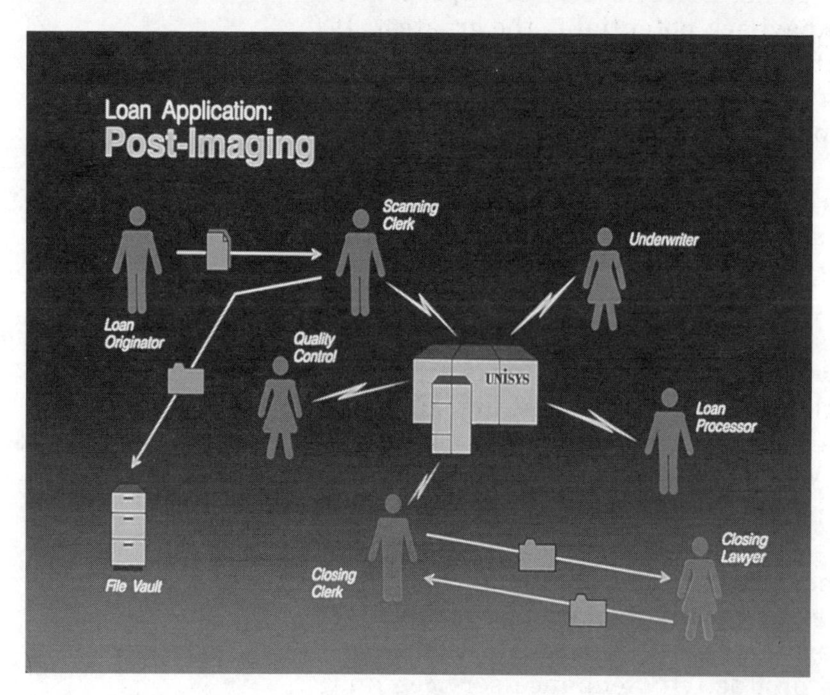

- **Look for revenue-producing options.** Like re-engineering, imaging is an opportunity to add value to government services. By charging an appropriate fee for the new value, local governments can create a revenue stream where none existed before. Most often the value comes in the form of expedited services, for which many businesses are willing to pay more rather than spend time doing business the traditional way.
- **Follow an implementation plan.** The phases of an implementation plan allow local governments to establish a clear set of tasks and responsibilities within an understandable decision-making framework. Breaking the tasks down into identifiable and manageable steps helps to keep the imaging project on track. The phases also serve as appropriate check and review points for management. Because of the level of detail involved in imaging implementation, this oversight and review can be critical in keeping the process from bogging down in minor issues.
- **Gauge the impact of imaging on the existing computer systems.** These days, imaging is rarely implemented as a stand-alone system. It delivers its greatest impact as an integrated system. This often means linking document images with existing databases and transporting images across existing networks. Potential users of imaging need to be certain that the imaging application can integrate as seamlessly as possible with existing databases and that it will not present a "traffic" problem on a government network.

- **View imaging as an entry point into a new field of computing.** For many local governments, imaging is the first foray into the world of client/server computing, which is the predominant computing platform for imaging. Client/server systems are much more versatile and cost-effective than mainframe computer systems. If well-planned, an imaging system can help a local government eventually migrate from its mainframe system to a less costly client/server system.

---

[1] Association for Information and Image Management, *Information and Image Management: The State of the Industry 1991* (Silver Spring, Md.: AIIM, 1991).

[2] Gartner Group study.

[3] Harold F. Langworth, "Imaging Capabilities in the 21st Century," in *Technology 2001: The Future of Computing and Communications* (Cambridge, Mass.: MIT Press, 1991), pp. 34-48.

[4] Association for Information and Image Management, *Information and Image Management: The State of the Industry 1991*.

Chapter 2

2 Policy Issues

Introduction

It may be the line workers in an organization who most need imaging, but the parameters within which the technology can operate must always be set at the top. An imaging system raises a number of issues for management. Without clear policy guidelines for management to follow, these issues can become serious problems once the imaging system is in operation.

To begin with, every local government needs to guarantee that storing documents on an optical imaging system will ensure that long-term access needs are met. Retention schedules need to be formulated and any legal issues pertaining to the admissibility of document images in court must be settled. At the same time, city and county leaders need to establish policy concerning public access. By placing a computer connected to an imaging system on a service countertop, government makes its records much more readily accessible than ever before. Policy is needed to balance the right of public access with the right of privacy.

Policy leaders must also determine the system's province and ownership. Will the imaging system serve the needs of a workgroup, or will it encompass an entire department? If the system serves several divisions of a department, who controls the system? And how does a government balance the strategic need for cross-departmental access to document images with the high costs of providing that access? In other words, where does the compromise occur?

Local leaders also need to develop policy for financing options. While the cost of individual hardware and software products continues to fall, the overall cost of planning and implementing a medium- to large-scale imaging system still remains high. Cities or counties in need of imaging but lacking the funds might wish to explore alternative financing options. To do that, however, their policy leaders must point the way.

Finally, leadership is needed to decide whether or not to generate revenue from imaging by adding a fee-based service. In numerous cases, an imaging system has enabled a department to provide a value-added service based on fast, electronic access to documents. It takes qualified leadership to judge whether revenue-based service is the right route to take.

Retention Rules

A government policy concerning document retention must be in place before imaging can take firm hold. Imaging solves an age-old problem—document storage—with an entirely new medium—optical disks. However, without a sound understanding of optical imaging and rules to govern how the technology is used to retain documents over the long term, users can leave the door open to potential problems.

According to a joint report by the National Archives and Records Administration and the National Association of Government Archives and Records Administrators on imaging guidelines for state and local government agencies, "long-term usability of digitally stored information . . . will be achieved by implementing a sound policy for migrating data to future technology generations." The report goes on to say that "ensuring the quality of digital images means exercising continuous control over three processes: conversion of the original image to digital data; enhancement of the digital image; and compression/decompression of the digital data for storage and retrieval."[1]

Other recommendations include:

- Establish retention schedules that specify the length of time that the records are needed to support program delivery and decision-making.
- Hold the agency accountable pursuant to law and regulation.
- Ensure the quality of digital images captured through an electronic conversion process.
- Provide for the continuing functionality of system hardware and software components over time.
- Limit the deterioration of optical storage media.
- Anticipate technological developments and plan accordingly.[2]

In Minnesota, statutes require the Minnesota Historical Society to specify standards for the reproduction of government records. These standards, which have been expanded to cover imaging systems, include system selection, systems administration (including public access and privacy, risk management and legal admissibility), and technical issues (storage media, scanning, quality assurance, image file headers, compression, documentation, storage, security and magnetic media).

In Texas, the state has set standards and procedures for electronic records of local governments through its administrative code. Among its many points, the code specifies that "data maintained on optical disks must be recopied a minimum of once every 10 years."

In addition, it specifies that the following standards be met for electronic records stored as digital images on optical media:

- A non-proprietary image file header label must be used, or the system developer must provide a bridge to a non-proprietary image file header label. Otherwise, the system developer must supply a detailed definition of image file header label structure.
- The system hardware and/or software must provide a quality

assurance capability that verifies information that is written to the optical media.
- Periodic maintenance of optical data storage systems is required, including an annual recalibration of the optical drives.
- Scanner quality must be evaluated.
- A visual quality control evaluation must be performed for each scanned image and related index data.
- A scanning density with a minimum of 200 dots per inch is required for recording documents that contain no type font smaller than six points.
- A scanning density with a minimum of 300 dots per inch is required for engineering drawings, maps and other documents with background detail.

Legal Issues

After cost, legal admissibility is the second most frequently considered optical storage issue.[3] According to Robert F. Williams, president of Cohasset Associates, Inc., a management consulting firm, all 50 states have laws that provide a solid legal foundation for admitting optically stored records in court. He adds that "although the specific legal basis may differ from one state to another, in most states it is found in the Rules of Evidence."

Documentary evidence is covered by the Uniform Rules of Evidence, which have been adopted in their entirety by the federal government and in some form by at least 28 states. The rules deal with duplicate and magnetically stored documents.[4] Those of the state of Minnesota, for example, state that "government records stored in imaging systems that meet standards set by the state historical society, are deemed for all purposes to be the original records."

Documents stored on optical disks are not automatically admissible in court, and the responsibility for making sure that they will be rests with management, which must develop procedures and guidelines for their organization.[5] Donald Skupsky, president of the Information Clearinghouse, says there are three requirements for authenticating the acceptability of electronic records in a court of law:

1. The existence of written procedures describing what the electronic imaging system is supposed to do

2. Documentation of training given to the operators of the system

3. Procedures for auditing the accuracy of the images.[6]

But according to Williams, optically stored records, as well as microfilmed and magnetically stored records, need to meet three additional requirements: accuracy, reliability and trustworthiness. One way to meet these requirements is to document procedures from the capture or scanning stage through backup, recovery and indexing.[7]

To increase the likelihood that optically stored records will be accepted as reliable and accurate, and therefore trustworthy by courts, the Association for Information and Image Management offers the following guidelines:

- The records should be part of a regular activity.
- Methods, such as quality control and audits, should be used to ensure or enhance the accuracy of the records.
- The timeliness of the records will increase their likelihood of acceptance.[8]

Workgroup Imaging vs. Departmental Imaging

Local government's first exposure to imaging technology was fragmented, with benefits and services usually restricted to one section of a department. Most often the imaging system simply replaced a paper-based storage and retrieval system. While local governments continue to benefit from imaging systems that are aimed at managing document storage and retrieval, the trend is moving away from islands of automation and towards integrated systems that can serve several purposes and a much broader range of users.

Workgroup Imaging

Single-application imaging systems are often called workgroup or desktop imaging systems. They serve a single purpose and do not share information with other computer systems—neither sending images across networks to other users, nor receiving data for use with the imaging application.

The benefits of workgroup imaging are low cost and simplicity. Workgroup imaging systems are usually small and require only inexpensive hardware and software licenses. Since they don't require integration with other information systems, such as a database on an existing mainframe computer, they are often easy to use. The small number of users also simplifies the imaging system's networking needs, including storage space, data traffic and administrative responsibilities.

However, the lower costs of a single-application imaging system can be deceptive. Certainly, the system doesn't have to support dozens of users, but most of the basic components needed for a small system are the same as those needed for a larger one: scanner, file server, workstations, optical storage device, printer and network.

Client/Server Technology

Cost and growth limitations were the major drawbacks of early imaging systems. Mainframe and minicomputer systems, which ran the first imaging applications, consisted of a single large computer attached to many terminals. As more users were added, the system's capacity shrank, slowing down performance significantly. The only way to boost the power was to purchase a new computer, involving hundreds of thousands—even millions—of dollars.

Client/server systems are much more than a local-area network (LAN) of personal computers, which typically just share files and peripherals such as printers. A client/server system divides the computing tasks of an application, such as imaging, so that the workload is better distributed. Thanks to improved workload distribution, client/server systems process information faster, accommodate additional users with no reduction in performance and allow other applications (such as faxing or even a geographic information system) to be integrated without substantial recoding of the existing imaging application.

With today's client/server systems, imaging can grow flexibly and cost-effectively. Since a client/server system consists of a network of low-cost personal computers or workstation computers, expansion means adding another computer and more performance to the system, not just another user. As a result, today's imaging systems have what is known as scalability: customers start with the basic components and add more users and features incrementally, without straining future budgets. A single, self-contained imaging system can't deliver the same advantage.

A greater disadvantage of self-contained imaging systems is their inability to share information with other users and systems. A storage and retrieval system might alleviate incredible pressure in an office that has been drowning in paper, but it doesn't add value to the department's overall information management. For reasons of fiscal austerity and improved governance, office or departmental information systems can no longer operate in a vacuum. Information needs to be shared today more than ever, and there's no better way to do that than electronically.

Departmental Imaging

Imaging has made what was once virtually impossible, possible. Government documents from one department can be viewed by another in seconds. These document images can be searched and accessed individually, or, with workflow tools, can be automatically routed to the proper individual on a regular basis.

Consider an imaging system used to manage crime and accident records for a law enforcement agency. By extending the system to the district attorney's office and to the local courts, important document images and the data attached to them, such as names, addresses and other vital statistics, can be transferred automatically to attorneys, court clerks, even judges. Not only does imaging create a single point of data entry for the entire judicial system, but it also provides a much greater information "picture" for the attorneys and judges who must hear cases and judge their merit.

Other departmental or multi-departmental imaging systems can bring similar benefits. Planning departments can share document images with building inspectors and engineers. Assessors, city auditors, title companies and attorneys can share land records.

Building these larger departmental or cross-departmental systems is not easy. A number of key issues must be weighed before venturing forth. Technologically, these systems are more complicated to run, often requiring an overhaul of existing networks to allow for rapid transmission of images without adverse effects on existing data transmissions—all of which means that costs can rise considerably. Issues concerning document-sharing across office or departmental boundaries, such as security and rights of access, have to be addressed. The impact of imaging on the work process of many employees accustomed to paper-based tasks must also be assessed.

Ownership and Control

On the surface, ownership and control of a workgroup imaging system might appear more straightforward than for an interdepartmental system. One might assume the smaller system is owned and controlled by the users, while the larger system spreads ownership and control among many. In reality, however, the issues are complicated, no matter the size of the system. Ultimately, the users should be in control, but of what? The hardware? The software? The work process? And who has overall ownership of the system? The users or the management information systems (MIS) department? And when something goes wrong, who gets called to fix the problem?

One approach to deciding ownership and control with an imaging system might be to follow the analogy of a leased car. It's owned by one person or business but controlled by another. In imaging, the system's infrastructure—hardware, network, operating system—is owned by the information systems (IS) department, while the imaging application is controlled by the users.

Sometimes, however, it isn't that straightforward. Imaging today runs on client/server technology, where the application and even the database can be split between several computers. Trying to decide where ownership and control fall gets more complicated, especially if the application is spread over more than one department, or if more than one application runs on the computer system. One approach to this problem might allow the IS department to set overall standards and objectives for the computer system and the user departments to control and own everything else.

Deciding ownership and control should be the responsibility of the imaging task force or project team. The team should make its decisions based on the needs of the organization. Ownership and control of the major components of imaging systems should be settled early. Some of these components include:

- Standards
- Schedules
- Budget and costs
- Hardware support and maintenance
- Application software
- Training
- Project management
- Infrastructure (hardware, networks).

Funding

Computer technology has advanced rapidly in recent years, driving down hardware and software costs while boosting performance manyfold. As a result, entry-level costs for a well-equipped imaging system have dropped dramatically. That's good news for small governments or governments with small-scale imaging needs.

Yet for many local governments, the first experience with imaging is usually costlier than anticipated. Often the department that needs the technology most has the biggest backlog of paper documents for conversion, lacks the kind of network necessary for distributing document images and needs new hardware to support the entire application. In addition, any imaging project involves extensive management costs—for everything from cost-benefit analysis and re-engineering studies before the system is built to training and maintenance during and after installation.

These non-technical costs need to be planned for extensively, since they can far exceed the costs of hardware and software. Here's a rule of thumb: five percent of the total system cost is for hardware, 15 percent for software and the rest—80 percent—for planning, management and training.

Imaging can be funded as a standard budget allocation, from a capital improvement plan, revenue bonds or a variety of sources.

Funding Ideas—Austin, Texas

- Enterprise departments can budget for a system.
- General fund departments may be able to budget for a system.
- Some areas may be able to use quality council funding.
- The city could form a partnership with a vendor.
- Departments could generate revenue through their imaging-related products and then use the funds for future imaging technology.
- The city could set up an enterprise-wide fund so that document imaging purchases could be combined. This method would encourage standardization and economy of purchasing.
- The city could develop partnerships with other government entities.
- The city could change the budget process for Capital Improvement Plan (CIP) fund purchases so that a percentage of all CIP technology purchases (for example, one percent) would be set aside for the city-wide records management program now under way. This program is mandated by [Texas] state law, and document imaging is one way to fulfill some of the records storage and retention requirements.

Source: "Recommendations for a Document Imaging System," City of Austin, Texas.

Numerous law enforcement imaging systems have been funded by drug asset seizures.

Non-traditional sources of funding have become common as fiscal austerity has forced many departments to be more creative. One option, lease-purchasing, has been in use by larger government entities for some time, but is beginning to be used more frequently by smaller jurisdictions. Lease-purchase plans offer the advantage of more affordable payments spread out over a specified length of time. A well-developed lease-purchase plan can also provide the local government with support services and perhaps a provision to upgrade the technology at a discount.[9]

A relatively new option is financing through cost-saving paybacks. Systems integrators—companies that design and build large, integrated computer systems for others—are beginning to use this approach to help governments finance the high cost of large-scale systems. When city or county governments carefully choose an application that generates measurable cost savings through automation, they can pay for the system through the savings, without any up-front payment.

A third funding option is to operate an imaging system as a service bureau for other departments, jurisdictions or organizations. Costs could be shared through a joint or regional partnership, much like some large-scale geographic information system partnerships that have been formed to share costs—and data—between neighboring jurisdictions, as well as public- and private-sector organizations.[10]

Public Access

Many jurisdictions offer public access to information in government databases. Often the right to access comes at a fee; in other cases, public access is free. Imaging systems represent a potential way to significantly broaden public access to government records in document form. Documents once available only to one person at a time and only within the vicinity of that person's location (i.e., the filing cabinet located in a downtown office building) can now be seen by many people simultaneously, whether they are in the same room as the imaging system, down the hall or across town.

Currently, direct public access to documents via an imaging system remains limited. Direct access allows the public to search and retrieve actual images at a terminal; indirect access allows users to view only an index of available documents. In 1991, a survey of operational imaging systems in government found only four out of 60 that permitted direct public access to documents.[11] Since then, these numbers have certainly grown, but the fact remains, most government departments with imaging are cautious about offering direct public access.

One reason for limiting direct access is network capacity, which can be affected by both the number of imaging users and the quantity of document images being retrieved at any one time. If a department were to install a number of public access terminals without any guidelines or controls, it may soon find its computer network crashing to a halt as public and staff attempt to retrieve large numbers of document images simultaneously.

Other reasons for limiting direct access are security and confidentiality. Until all scanned documents have been cleared for security purposes, a department will probably first want to err on the side of security rather than give the public full access to documents stored on optical disks. If the imaging system contains public records as well as material of a sensitive nature, then it makes sense to investigate methods for limiting access without cutting it off entirely.

Most imaging systems come with some kind of password feature, which can prevent intentional or unintentional viewing of documents or alteration of an index. The more versatile imaging software systems have toolkits that programmers can use to develop highly customized security and password features that can be controlled by department administrators.

Commercialization

A strong reason why public access needs to be reviewed at the policy level is the potential for generating revenue. Just as city and county governments have found on-line database records a potentially lucrative source of revenue, they are discovering that imaging systems can be configured to provide valuable services for which businesses are willing to pay. Raising revenue through the sale of data and information is not new. In a 1992 survey by the Syracuse University School of Information Studies, 60 percent of responding county governments nationwide said that they sold data in some form.[12]

Imaging is seen by many as particularly well suited to revenue-generating applications because it provides valuable access to vital documents. Businesses, in particular, are often willing to pay a higher fee for timely access to documents. Local governments justify the higher fee by pointing out that it forces the relatively small group of customers who use the service to pay for it, freeing government from having to fund the select group's access through taxes.

A case in point is land records. While every property owner must file these documents to establish legal ownership of his or her home or land, government invests a lot of taxpayer money in making the documents available to a small group of people, primarily attorneys, title companies and individuals involved in title searches and other matters concerning land records and deeds.

Imaging has been introduced at a number of county registry offices to reduce storage space problems, improve document retrieval and expand access. Some innovative registry offices have turned the latter benefit into a value-added service, whereby businesses, for a fee, can access the database (not the images) of land records available on optical disk. Through such features as fax gateways, these businesses can have copies of documents sent directly to their offices, without ever having to enter the county registry building.

Revenue Risks

A revenue-generating imaging program can be profitable. But with the potential for profit comes the risk of failure. For many agencies, the ramifications of failure—lost taxpayer funds, political fallout—are enough to stifle any attempts at dabbling in the commercialization of document services. Also weighing against the success of profit-making in the public sector are the built-in restraints of precise rules and regulations that city and county workers must follow.

If public-sector entrepreneurship can survive these obstacles, there is still the issue of seed money with which to launch the project. In the private sector, a business can sell shares or try to attract contributions from venture capitalists. But in the public sector, finding money is an entirely different problem. According to Stuart Bretschneider, director of the Center for Information Policy Studies at Syracuse University, finding the funds for such a venture "typically becomes an entrepreneurial shifting of funds" from an existing program to the proposed profit-making venture.[13]

Marketing is another profit-making concept that's alien to the public sector. Service-oriented government managers, trained from day one to think in terms of the "public good," must learn the art of sales and marketing to ensure success for any profit-making venture.

Marketing begins with research to determine the customer base, its location, its size, and the amount of money it is willing to spend for the service the department plans to offer. Once the market has been defined, the department needs to inform targeted consumers about the service and entice them to use it. Direct-mail and advertising are two major avenues for reaching the customer base.

While plenty of books and experts explain the do's and don'ts of marketing and advertising to the private sector, no such resources exist for public-sector entrepreneurs seeking advice. However, a number of innovative local governments have succeeded in profiting from their information technology, and Public Technology, Inc. (PTI), has researched public enterprise and advised local governments on its application for years.

Many people still consider imaging an evolving technology that has yet to mature. For some people, adding a risky profit-making component to an imaging application might seem the height of foolishness. Yet, a small but growing number of local governments are using their imaging systems to generate revenue. The concept does work if implemented properly.

---

[1] National Archives and Records Administration and National Association of Government Archives and Records Administrators, *Digital Imaging and Optical Storage Systems: Guidelines for State and Local Government Agencies* (Washington, D.C.: NARA, 1991), p. 14.

[2] Ibid., p. 15.

[3] Robert F. Williams, "Is It Legal?" *Document Image Automation*, vol. 12, no. 3, p. 73.

[4] Brian Miller, "Imaged Documents and the Courts," *Government Technology*, March 1994, p. 34.

[5] Ibid.

[6] "Electronic Records Admissible as Evidence, Experts Say," *Government Imaging*, March-April 1994, p. 1.

[7] Miller, p. 56.

[8] Ibid.

[9] *The Government Executive Imaging Handbook* (Sacramento, Calif.: GT Publications, 1992), p. 13.

[10] Ibid.

[11] National Archives and Records Administration and National Association of Government Archives and Records Administrators, *Digital Imaging and Optical Storage Systems,* p. 21.

[12] Brian Miller, "Profits in Government?" *Government Technology,* February 1994, p. 42.

[13] Ibid.

3

Chapter 3

Management Issues

Introduction

Unlike other information system projects, imaging presents a unique challenge to management because of its far-reaching impact on an entire organization. With 90 percent of its existing information in the form of paper documents, government still works in much the same way it has for decades. With imaging, however, cities and counties may change the information status quo and gain significant new benefits in productivity, efficiency and service. But imaging also has the potential to bring about enormous change in strategies, processes, jobs and even culture.

To manage both the benefits and change that imaging brings, local government has at its disposal both old and new tools, techniques and methodologies that can ensure successful planning and implementation. Strategic planning, cost justification, needs assessment, requests for proposals and pilot tests are some of the traditional methods used by management to bring up an information system. Added to these are some new management tools, such as change management and business process re-engineering, which give local government managers the ability to harness the great capabilities of imaging.

Strategic Planning

So where does one start? Management should begin by strategically planning for the implementation of imaging. Strategic planning helps a local government determine whether or not imaging is a feasible solution for its document problems. Without a plan, the government or department that installs an imaging system is bound to see its project fail.

Strategic planning allows a local government to devise a blueprint of what its current needs are, what kind of benefits it is seeking, and how it is going to reach its goals. Local governments or specific departments begin by scanning their current environment to pinpoint the biggest bottlenecks in records management, workflow, networking, and database management systems.

With a strategic plan, a city or county can identify how imaging will be used, what its biggest risks will be, where it should be implemented first for maximum benefit, and how it will improve such things as customer service and worker productivity. Should conflicts arise during implementation, the vendor or systems integrator and the government department can refer to the plan to help everyone get back on course.

A strategic plan begins with input from all departments that would be affected by the imaging system. Users and management then identify opportunities and lay out objectives and goals for the long term. The plan spells out some of the major costs involved and the benefits that can be expected. Finally, it lays the groundwork for the next step in the process: the feasibility study.

How necessary is a strategic plan? Because of the complexity of imaging technology and the amount of change it introduces to an organization's work process, the overwhelming consensus is that every local government needs a strategic plan to ensure the success of an imaging project. The best approach is to establish an overall city- or county-wide plan, but strategic plans specific to a department are acceptable. The reason most imaging systems fail in local government can be traced to the lack of a cohesive plan for users and vendors to follow.

Hennepin County, Minn., developed a strategic plan to "assist decision-makers who are planning and implementing imaging systems in Hennepin County over the next three to five years." The plan begins by covering the basics, including the main public records problems the county faces; how imaging can offer a solution; the organizational and technical challenges that can derail any successful use of imaging; and the steps decision-makers need to take to successfully implement imaging.

An Imaging Task Force

How does a local government begin a strategic plan and start to examine the feasibility of imaging technology? A task force can be an effective way to pull together the diverse interests of a city or county government and merge them into a common mission with specific goals.

A task force, unlike a committee, is an action-oriented group, consisting of key personnel from all branches of government, with a clearly defined role and a specific time frame in which it must get its job done. The most effective task forces are those with top management support. The role of the task force is not to guide a project, but to advise city or county leaders on the problems, the kinds of opportunities that exist and how to go from there to a working solution.

With a task force, local governments can avoid the all-too-common problem of individual departments going off and building separate imaging systems. By setting a global view—through the strategic plan—the task force can establish parameters, guidelines, even basic standards that enable the implementation of imaging systems with maximum benefits.

Hennepin County set four strategic goals to guide the county's acquisition of imaging technology:

1. Determine general and application-specific county imaging requirements.

2. Document hardware and software platform assumptions and alternatives that may impact imaging.

3. Educate county management and staff.

4. Maintain a profile of installed imaging applications and plans for those anticipated within the next five years.[1]

The strategic plan goes on to elaborate on these four goals, and includes projections on the number of county government workstations that will be used for imaging.

Change Management

When a local government prepares a strategic plan for imaging, it should quickly see how dependent its current work applications are on paper and how many people would be affected by an imaging system, compared to other forms of data automation.

Clearly, change brought on by imaging can alter an organization's business, as well as its strategies, processes and culture. Change can also make employees skeptical, uncertain, even doubtful about the benefits of the technology. In fact, resistance to change should probably be one of the key risks of imaging identified by a strategic plan. Unfortunately, if the issues of change—as they affect both the people

Change Management and the Naysayer

What happens when you run into a manager who is dead set against imaging as a possible solution? Do you try to include that person in the project or exclude him or her from it?

Usually the naysayer has a legitimate concern that is camouflaged behind his or her criticism and negative attitude. Instead of isolating the naysayer, spend some time with the person, one-on-one. Find out how the imaging system will affect the manager. Tie the benefits of the system to overall government goals so he or she can see how it will work for the good of the entire organization and not just the few offices or departments directly involved with the system.

If you take the time to understand the manager's concerns and explain the universal benefits of the system, chances are he or she will end up supporting the project. And as any project manager knows, there is nothing better than a born-again believer.

and the organization—are not dealt with adequately, they can lead an imaging application to failure.

The process for helping a government address these problems is called change management. Government managers identify users' needs, fears and expectations and bring about their acceptance of change. To best handle change, a local government must make sure that three key activities take place:

- The changes and their ripple effects must be understood as fully as possible.
- Ongoing communication with the workforce must take place. Two-way dialogue will foster understanding of the changes and the ripple effects.
- Activities that help smooth the transition must be implemented.[2]

The two most common ripple effects of imaging are reductions in paper and changes in workflow. The services that employees provide will also change, and possibly the type of command and control used to run a work process.

A major concern to employees is the possibility of job loss resulting from changes in or elimination of certain tasks related to paper handling. Computer phobia may grip some employees. All of these ripple effects must be identified and discussed before they begin to dominate and undermine the project.

Once management understands the changes that imaging can produce, it needs to communicate those findings to staff. Job changes and possible elimination should be addressed directly. Changes in workflow and expectations should be discussed with staff. For example, employees used to waiting minutes to retrieve a paper document will suddenly find a 45-second wait at the computer to retrieve an image too long.

During the transition, management can offer training, ongoing communication and support, and job redefinition to help smooth system implementation.

Work Process Re-Engineering

In 1990, an article appeared in the *Harvard Business Review* that was to have a profound effect on management, information technology and, in particular, imaging. The article, "Reengineering Work: Don't Automate, Obliterate," by consultant Michael Hammer, explained that organizations, if they changed the way they acquired and used information, could boost productivity not by a few percentage points, but by hundreds.[3]

The concept, known as work process re-engineering (or simply re-engineering), has been embraced by private-sector firms and a growing number of public-sector institutions, including local governments. Re-engineering calls for a complete redesign of how work is performed and services are delivered. It starts by setting goals, then rethinks pathways for task completion, with technology as the change-enabler. The principles of re-engineering can be summarized as follows:

1. Organize around the outcome, not the tasks.
2. Have those who use the output of a process perform that process.
3. Integrate information processing with information production.
4. Treat geographically dispersed resources as though they were centralized.
5. Link parallel activities instead of integrating their results.
6. Let those who perform the work make the decisions and build control into the work process.
7. Capture information once at the source.

Business process re-engineering is one of the key benefits of imaging. It ensures that a local government does not pave the cowpath with automation, but actually redesigns the business process to create significant improvements in cost, time and value. For example, imaging allows an organization to capture information at the source and centralize its access. Incoming paper documents, once treated as

A Re-Engineering Crib Sheet

What Is Re-Engineering?

- Fundamental redesign of work processes
- Radical and rapid performance improvement
- A sea change enabled by information technology

What Should Be Re-Engineered?

- Customer service (one-stop government)
- Cross-program integration (sharing information across turf boundaries)
- Overhead and administration (flatter organizations, empowered workers)

When and How to Re-Engineer?

- When outside support can be mobilized
- Slow trigger, fast bullet
- Entrepreneurial start-up

How Do I Get Ready?

- Develop a vision, a plan and an infrastructure
- Educate managers
- Find creative sources of funding
- Measure performance and make results visible

Source: Jerry Mechling, "Reengineering: Part of Your Game Plan?" *Governing*, February 1994, p. 48.

separate sources of information, can be scanned and indexed with existing databases of information, so that employees can locate and access *every* piece of information—whether it concerns a crime, a parcel of property or an individual's health record—at a single source, by pushing just a few buttons on a computer.

Imaging also allows an organization to turn sequential or serial work processes into parallel ones. Whereas in the past each staff person had to finish his or her paperwork before passing it on to another, imaging systems now enable various staff to view document images simultaneously, so work is processed by several people at the same time. Because imaging removes paper from a work process, it allows an organization to remove the "buffers" and "hand-offs" that can slow down transactions and other processes.[4]

Process redesign achieved through imaging can slash both the time it takes to deliver a service and the cost of delivering it. Re-engineering also helps an organization wring value from its redesigned process. Think how valuable a re-engineered work process would be that allows a title company to search and fax to itself copies of deeds without leaving the office, rather than going downtown to request that a government clerk perform the same task. Clearly, re-engineering can save time and money for both service providers and service recipients.

Synergy of Business Process Redesign and Systems Renewal

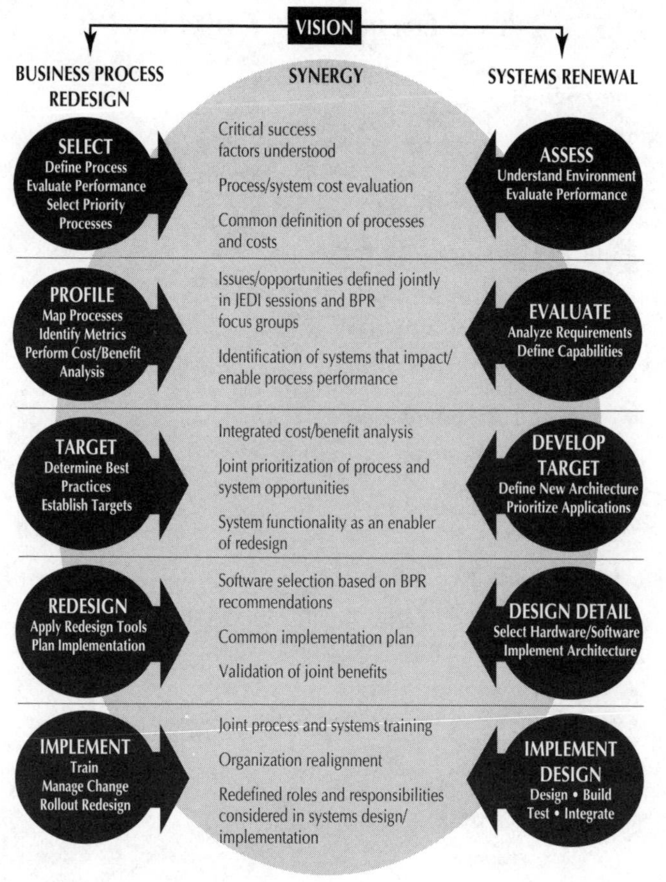

Source: Booz-Allen & Hamilton

But planning and implementing imaging systems that re-engineer the flow of work in an organization can be extremely challenging. Imaging systems provide a broad range of options for supporting and enhancing business operations. The interplay between what already exists in an organization and what is envisioned makes implementing imaging systems that support re-engineering time-consuming and complex.[5] Some local governments rely on consultants or systems integrators to help guide them through the process of re-engineering. Others have brought the skills of re-engineering in-house and used them again and again as imaging applications proliferated.

If re-engineering is passed over for lack of planning, the results can be disastrous. John Bennett, executive director of Connecticut's Office of Information and Technology, describes what happened when two state agencies tried to implement imaging without redesigning their basic work operations. "They tried to turn a paper process into an electronic process," he says. "It got very expensive very fast and they got a very poor response very fast. The net results were not really that much better than what they had before they started."[6]

Government agencies that have implemented imaging are warning others not to underestimate the scope of work involved in re-engineering and change, especially for large-scale imaging systems. Tom Carroll, project director for imaging systems at the state of Washington's Department of Labor and Industries, says that change management alone (which, in his definition, includes management of re-engineering and other processes), can constitute one-half the cost and time an imaging project demands.

Carroll's department has implemented a large imaging system to process unemployment compensation claims. "I started out thinking that imaging was one-half systems issues and the other half applications issues. I soon changed the percentage to one-third systems, one-third applications and one-third change management. I finally realized that the percentage of work was one-half systems and applications issues and one-half change management issues."[7]

To help government imaging customers avoid the kinds of mistakes that have occurred in Connecticut and elsewhere, many vendors—imaging system providers as well as general systems integrators and computing consultants—are offering methodologies, tools and services that assist the re-engineering process. A successful re-engineering project, however, begins with the department itself setting a clear and simple goal for the project—not just a slogan. Here are several other ingredients in the recipe for success:

- **Don't let technology drive the re-engineering process.** Re-engineering isn't about new technology—it's about work design and organizational change. Technology is just an enabler.
- **Build front-line employee support.** Good leadership is essential to ensure that the difficult task of re-engineering takes place, but front-line employee support is crucial. With most staff responsibilities reassigned through re-engineering, some individuals are bound to resist the change. Try to minimize the resistance by getting these individuals' buy-in on both re-engineering and imaging before change actually begins.

County Records—Imaged and Re-Engineered

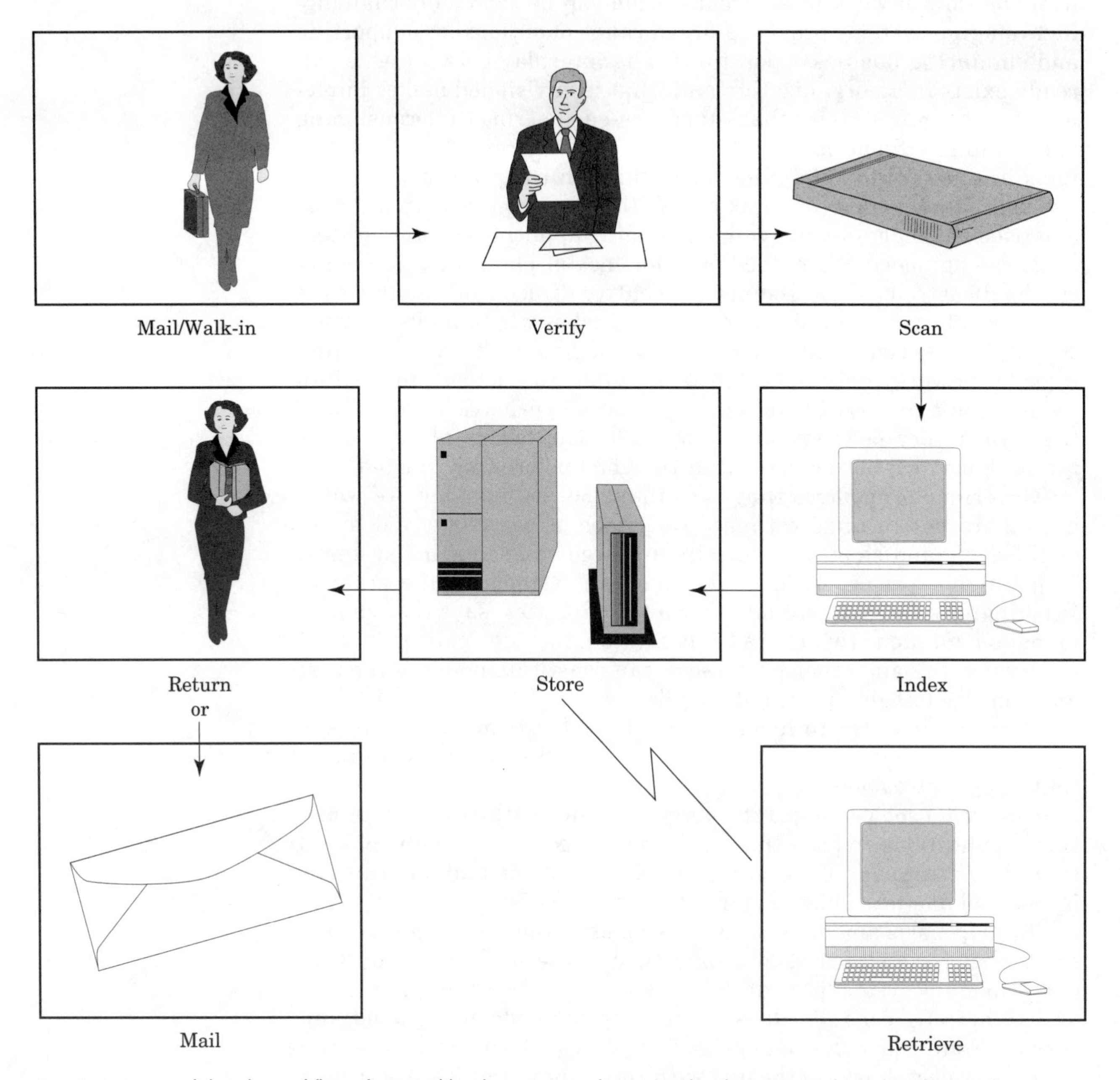

Imaging can consolidate the workflow of a typical local government department by eliminating tasks such as filing and photocopying and other steps that slow down transactions. The diagrams on this and the following pages illustrate the process for handling county records before and after re-engineering and imaging.

Source: Unisys Corporation

County Records—Pre-Imaging

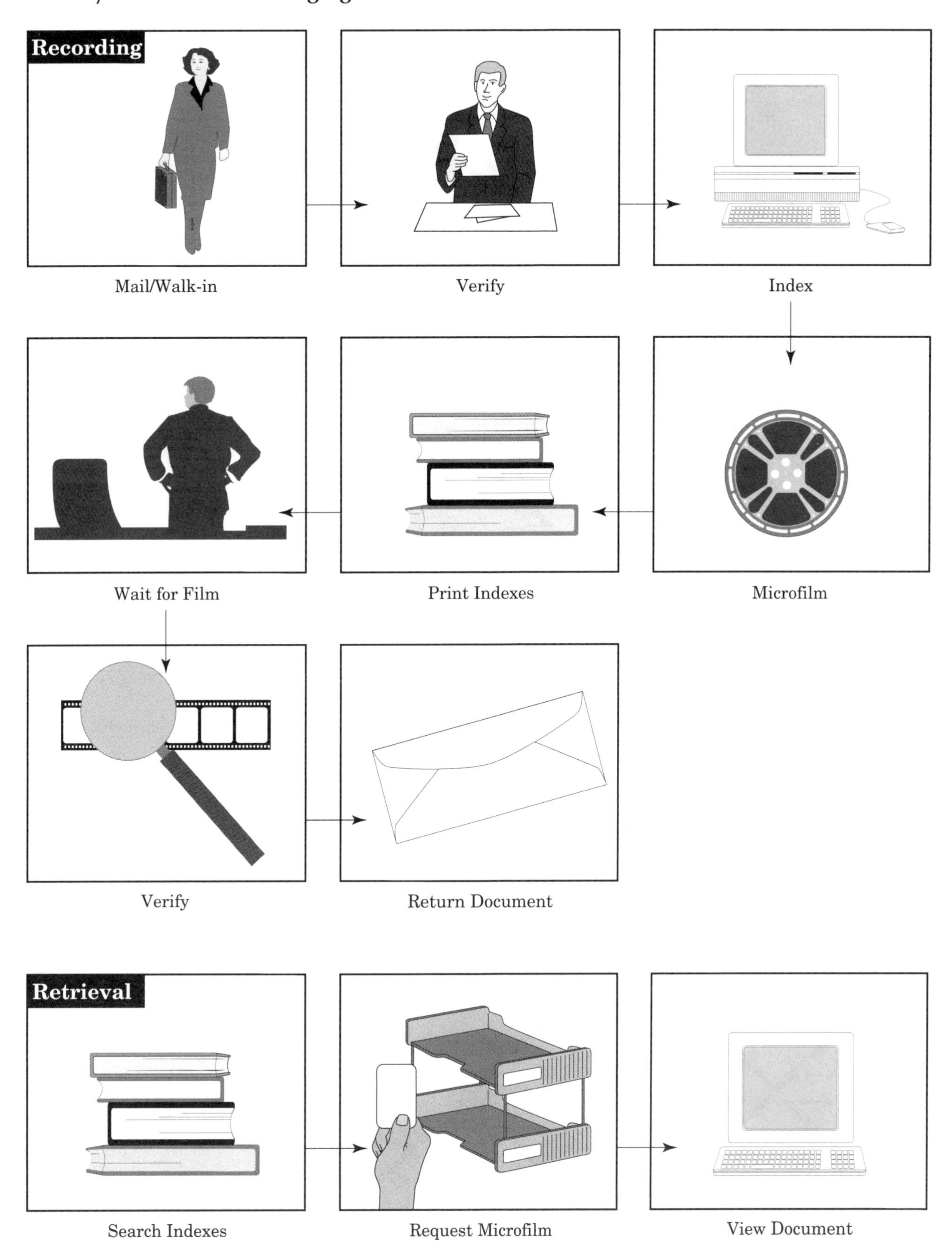

County Records—Post-Imaging

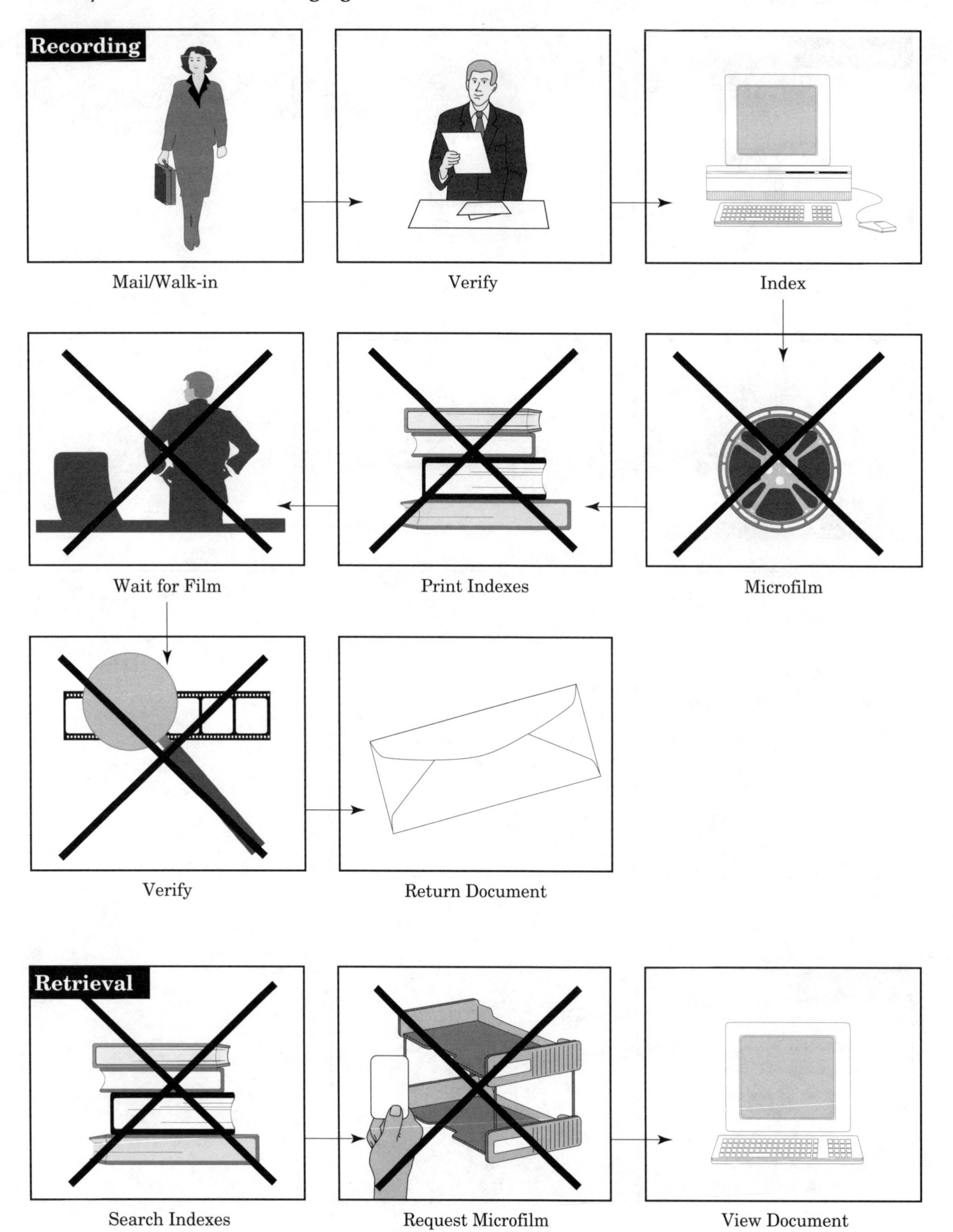

- **Keep expectations realistic.** When it comes to technology, overly ambitious designs can lead to delivery shortfalls, frustration and significant expense associated with reconfiguring the information system to satisfy reduced goals.
- **Analyze all aspects of operations.** A surprising number of organizations are incapable of mapping current work processes. As a result, management must take the time to analyze and define every existing process before re-engineering can take place.
- **Don't search for the perfect system.** Excessive deliberation can lead to a failure to commit. An imaging system in a re-engineered work environment doesn't deliver its payback until it is implemented.

Cost Justification

Price and performance trends in computing have had a significant impact on the cost of imaging systems. Just a few years ago, the average cost of a top-of-the-line imaging system ranged from $35,000 to $50,000 per user. Today, in many cases, those costs have dropped below $10,000 per user. (The figure can go higher, however, depending on the size and complexity of the application.)

Combine these price drops with significant performance improvements in computers, scanners, optical disk storage systems and local-area networks, and purchasing an imaging system has never looked better. But how much is imaging really going to cost? And is the cost justified by the benefits?

For local governments, the answer is best found through cost justification. Management typically begins cost justification by comparing the present cost of doing business with the estimated costs

Key Elements of a Successful Cost-Justification Process for Imaging

- Quantify expectations and develop a measurement plan for each expectation.
- Focus on the following costs and benefits:
  - Personnel costs
  - Equipment costs
  - Filing space costs
  - Benefits from expediting a process
  - Elimination of lost files
  - Benefits of distributing the process.
- Track costs and benefits over time.
- Measure results.
- Communicate benefits to management.

Selected Management Guides and Standards in Cost Expenditures

	October 1975	March 1980	February 1984	February 1988	June 1991
Cost of creating an average business letter	$3.31	$6.43	$7.69	$9.14	$10.26
Cost of filing supplies (per file drawer)	$48.00	$128.62	$153.36	$176.50	$197.98
Cost of cabinet (based on 10-year depreciation)	$10.00	$24.64	$30.10	$34.64	$38.86
Cost of office space for each four-drawer file cabinet	$30.00	$67.88	$80.82	$93.02	$104.35
Monthly cost of pay for one file clerk for 12 cabinets (based on 22 working days per month)	$352.00 ($80.00/wk)	$731.24 ($166.19/wk)	$882.98 ($220.74/wk)	$1,016.27 ($254.07/wk)	$1,135.13 ($283.78/wk)
Cost of overhead (25 percent of labor)	$88.00	$182.81	$226.50	$260.69	$292.40
Annual cost of owning and operating a standard four-drawer file cabinet	$528.00	$1,134.98	$1,272.76	$1,581.13	$1,768.72
Annual cost to maintain a standard four-drawer file cabinet (based on one operator for 12 cabinets)	$687.88	$1,028.85	$1,143.05	$1,361.50	$1,527.18
Estimated cost of filing (per inch)	$6.87	$10.29	$11.14	$13.26	$14.88
Cost of records stored in active files (per file drawer)	$4,965.00	$7,986.96	$8,761.42	$10,328.65	$11,585.53
Dollar worth of records in one four-drawer file cabinet	$19,860.00	$30,947.85	$35,485.68	$42,267.55	$46,342.09
Average cost of a misfiled record	$74.15	$96.45	$106.45	$126.81	$142.25

Source: The Records and Retrieval Report, April 1992.

under the new system, and then comparing the benefits under each method of operation. To measure the cost of imaging, management must analyze costs for hardware, software, application development, training, and such variables as back-file conversion, labor for scanning and indexing, planning and re-engineering, all of which can add time and expense to the project.

The quantifiable benefits of imaging include cost reductions and possible revenues from access fees. Other, not so quantifiable benefits include departmental enhancements, such as improved worker productivity and more efficient work processes, and strategic advantages, such as multi-departmental document-sharing and improved service delivery.

Measuring the Benefits

Local governments can begin to measure benefits by examining some of the costs associated with storing, maintaining and processing paper records. The annual cost of owning and operating a standard four-drawer file cabinet in 1991 was estimated at $1,768.72.[8] American business annually misplaces an estimated three to five percent of its records. The cost per misfiled record stands at approximately $200. Other paper costs include those for storage space ($10 to $30 per square foot), photocopying (3.5 cents per copy) and microfilm (four cents per frame). Imaging can have a sizeable impact on all of these costs, reducing them as well as associated labor expenses. These cost benefits are easy to identify and capture for analysis.

If local governments charge fees for access to imaged documents, they can quantify "hard" revenue benefits as well. While not every imaging application is a potential breadwinner, some areas of government operation present good opportunities for revenue generation through so-called "priority service" fees for access to on-line records and document images.

Harder to measure are the so-called "soft" benefits imaging can deliver. Operational enhancements that result from reducing or eliminating paper from the work process are a clear example. Another is simultaneous access to an image by many users. Both of these benefits, derived when re-engineering is combined with imaging, can boost productivity and cut labor costs. It is difficult, however, for management to quantify just how many tasks can be eliminated, how many hours of labor can be saved and how rapidly work-cycle time can be reduced. These unknowns may be hard to forecast, but with good up-front evaluation, based on solid planning and design, solid benefits are a certainty.

Cost Considerations

When broken down, the cost of implementing an imaging system corresponds to the three key components of an information system: hardware, software and what PTI calls "orgware," which includes re-engineering, training, maintenance and management associated with the information technology. Orgware also covers services for implementation, communications and integration, all of which have grown in importance as information systems become more complex to design and implement.

Imaging applications in local government, however, often entail another major cost component: back-file conversion of existing documents. For departments, such as land records, where access to archived documents is critical, back-file conversion is a must. The city of Scottsdale, Ariz., estimates that the cost of converting one box of documents to optical disk is approximately $1,200. These costs can vary according to location and the entity that sets the price (usually a service bureau).

Back-file conversion costs are often substantial and can easily exceed the cost of the imaging system itself. Sometimes, a good purging of the files can help reduce the paper volume and the cost of a back-file conversion. Palo Alto, Calif., asked its user departments to develop a list of key documents that had to be scanned for imaging. Then the city used college students to clear files of all unnecessary documents. Other local governments have found that back files are often loaded with several copies of the same document, each stashed in a different place.

Management must remember that while imaging can reduce some traditional labor costs, it also adds some new labor requirements.

File Conversion: To Outsource or Not

Before fine-tuning scanner choices, any government department considering an imaging system must evaluate its document conversion requirements and the strategy to follow. Some applications begin when the system is turned on and the first document is scanned into the system. This is known as day-forward conversion, with the system building its database of document images with each new day's incoming documents.

But for many other applications, the imaging system isn't operational until all documents in filing cabinets have been scanned and stored. Land records and police fingerprint cards are two instances where back-file conversion of some or all documents is necessary before the imaging system can be operational.

For those applications needing back-file conversion, local governments must weigh two options: converting the documents in-house versus using a service bureau to do so. Each option presents advantages and disadvantages. Service bureaus have high-speed equipment and trained staff to get the job done on time. They can also reduce a government's capital investment and keep conversion costs predictable. On the other hand, in-house conversion has the advantage of control; staff know what to weed from the document collection. If properly planned, in-house conversion can cost less than conversion through a service bureau.

Scanning and indexing, two tasks not performed in the paper environment, are required for any imaging system to function. With early imaging systems, many organizations reported that any labor savings resulting from reduced clerical work were often canceled out because of these added requirements. However, with improvements in optical character recognition (OCR) technology and redesigned documents that include bar codes, some organizations have been able to automate some, if not most, document indexing and have reduced the amount of labor needed for this task. (See Chapter 4, "Quality Control," for further discussion.)

Imaging may also require more advanced—and more costly—computer networks. Most existing local-area networks have the capacity to handle an imaging application—that is, up to a certain point. Should the number of users and the amount of imaging traffic increase significantly, just about any network will feel the impact, mainly in the form of severe slowdowns in data transfer for imaging users and non-users alike. Upgrading a network to a higher bandwidth to accommodate the growth can be a significant cost factor.

Other costs unique to imaging include the possible need to use parallel media (paper and image) until the new system is fully functional, and the reclassification of personnel whose jobs and responsibilities have changed through re-engineering and the introduction of imaging.

Outsourcing

When the city of Chicago decided to overhaul its poorly functioning parking adjudication program, it outsourced the project entirely, allowing the winner of the bid to decide what technology should be used and where. The outsourcer built a distributed imaging system, using its own computer hardware, software and networks to allow

Imaging Cost Considerations

- Imaging system (hardware, software, maintenance and operations)
- Network
- Integration with other applications and systems
- Back-file conversion
- Security and recovery
- Training
- Parallel operations
- Reclassifications of personnel
- Quality control
- Support
- Personnel termination costs

Sources: Unisys Corporation and Andersen Consulting

drivers to pay parking fines in their neighborhoods rather than drive into downtown Chicago.

As in most outsourcing projects, the city has a multi-year contract with the outsourcer and used performance clauses to ensure the project was built on time and according to specifications. Most local governments don't go the route that Chicago did, but outsourcing should not be ignored as an option, especially as a way to get a technology as complex as imaging up and running on time and within budget.

How much outsourcing is required depends on the skills and needs of the city or county. Most local governments will do a limited amount of outsourcing, usually in the form of contracting out application development. If imaging happens to be the first government application involving client/server technology, then it's possible the system will be planned and developed by a systems integrator with the necessary skills.

Tips and Recommendations for Identifying Cost Benefits

In his book *Implementing Electronic Imaging: A Management Perspective,* author Scott Wallace provides the following recommendations for identifying cost benefits:

- **Identify the primary objectives of the imaging project and make certain they align with the department's business and information plans.** If the existing network is already overloaded, then an imaging system will just make a bad situation worse. Cost/benefit analysis must include the price to upgrade the network, for example.
- **Identify costs of both current and proposed operations.** Be sure to compare the proposed system's costs with the *actual* costs of current operations. Remember that imaging not only reduces costs associated with paper storage, retrieval and distribution (i.e., photocopying), but also those associated with disaster recovery and time spent by staff returning calls because they didn't have the information available to answer a query the first time.
- **Identify benefits of the proposed imaging solution.** Beyond the benefits of reducing the amount of paper flowing through a department, imaging delivers value that can be measured in varying degrees. Don't overlook the "soft" benefits of enhancing information accuracy and integrity, improving customer service and enabling faster decision-making.
- **Assess, evaluate and recommend.** Cost justification and benefit analysis should provide management with the foundation to decide whether and how to proceed with the project. Although many imaging implementations have been justified on traditional cost-reduction terms, the most successful of these systems have more strategic primary goals: value benefits such as productivity, revenue and service. Management in these organizations wrestled with the analysis and became comfortable enough to proceed with a project yielding a mix of hard and soft benefits.[9]

Outsourcing has many benefits. Through outsourcing, local governments can tap the outside expertise needed to execute a project properly. They can also manage risk by setting deadlines and budgets that the outsourcer must meet. But local governments must remember that outsourcing does not mean giving up control or oversight of the project. Outsourcers don't have the same vision that local governments have. They need government leadership to ensure the project meets the goals and objectives of the city or county.

Steps in the Planning and Procurement Process

Once the general management issues of imaging have been identified, researched and settled by strategic planning, cost/benefit analysis and re-engineering, the actual planning and implementation can begin. The methodology applied here follows steps recommended by PTI's *Information Master Planning: A Guidebook for Local Government*.

The guidebook offers several rules of thumb to ensure that the plan remains effective: avoid excessive detail, address the right audience and foster action. When necessary, the guidebook suggests, the plan should be revised to reflect changes in organizational needs and technology.

Step 1: Obtaining Top Management Support

Unlike some technologies, such as geographic information systems, which are typically large and multi-departmental, imaging systems can be either small or large. Examples of both abound in local government. In general, however, the average imaging system in local government affects a fair-sized department and can cost from hundreds of thousands to millions of dollars.

At a minimum, the imaging system will have an impact on the tasks of numerous workers and the functions of the entire department. At the high end of the scale, an imaging system could affect several departments. Whatever the size of the system under consideration, it will require time, dedication and good management to plan and implement what many consider to be still an emerging technology. Top management leadership is essential in seeing that all requirements are met throughout the process.

Step 2: Establishing the Project Team

A team approach to planning and implementing an imaging system is highly recommended. The project team will negotiate work issues and implement the delivered system. If the imaging system crosses divisional or departmental barriers, the team can foster cooperation and participation among all involved, encouraging comprehensive solutions.

PTI recommends that jurisdictions adopt a two-tiered approach to imaging project management, forming two closely related teams: the policy team and the implementation or technical team. One individual, a project leader or coordinator, is charged with integrating imaging-specific policy issues. The role of the policy team is to define the key issues, allocate resources for planning and implementation, identify final goals and objectives and approve the implementation plan. The team should include the jurisdiction's top managers.

The technical team should include professionals from all departments or divisions with a stake in the imaging system. It is important to note that not all individuals participating in the team are required to have information systems or computer technology background. Often it is desirable to have a representative of top management chair the team, because his or her presence minimizes the political pressures on the project coordinator and frees the policy team from having to resolve minor issues.

Throughout the project, the project leader must have access to top management. He or she should be an effective advocate for change, understand the policy issues that the planning process will raise and have a good comprehension of information management.

To clarify the project's purpose and scope and to help team members recognize its central issues, some jurisdictions begin with a workshop on imaging technology and related planning issues. These introductory workshops are often very successful in establishing a common perspective.

Here are some issues specific to imaging that can affect the makeup and responsibilities of the technical team:

- If implementation of the imaging system involves integrating new hardware with existing hardware, then MIS departments will likely need technical help from vendors or system integrators. Today's imaging systems run on the client/server computing platform, which is relatively new to most city and county MIS departments. Often, imaging applications are integrated with existing databases that run on proprietary hardware. The technical team should avail itself of outside expertise to implement client/server technology. It should also involve MIS staff in the project to ensure that they understand the integration process.
- If the system involves significant software development, outside skills may well be needed. Again, imaging is an emerging technology and few, if any, imaging applications can run straight from the vendor's box. Customization is almost always required, especially for line-of-business applications that tackle a specific process.
- Spell out who will integrate software and hardware and perform system testing.
- If the implementation schedule is a fast one, then the technical team must identify critical paths and bottlenecks and develop workarounds and supplemental support when events threaten the schedule.
- No matter how extensive the collaboration with the vendor or integrator, the team will need staff experienced enough to know when processes are on track and capable enough to correct problems when they are not.[10]

Step 3: Assessing Needs

Needs assessment, a critical step in imaging planning, involves stating needs, analyzing problems and barriers and specifying long-term objectives in information management. The project team begins a needs assessment by detailing current practices. The goal is to create a comprehensive and diagrammatic description of how work is processed in the department. This analysis includes descriptions of all workflows and job functions, as well as the documentation and support necessary to perform work tasks.[11]

To compile this information, the project team needs to interview supervisors, to identify jobs and documents and to detail what hap-

Excerpts from Imaging Team Charter and Plan—Hennepin County, Minnesota

Formation and Purpose of the Team:

The general purpose or "mission" of an Imaging Team is to stay abreast of this exciting new technology, and in some cases, to recommend that it be considered as a solution for certain situations. Specifically, the purpose of the team is to provide departments with the needed project assistance or information, particularly if they are contemplating any move in the direction of new imaging systems.

Members, Roles:

The Imaging Team is informally structured. Representatives of various County agencies are invited to join as appropriate and approved by the Director of Information Services.

Schedule of Meetings:

The Team meets at least once a month.

Activities of Team:

1. Appraise potential imaging applications in County departments, prepare reports and recommendations, including cost estimates.
2. Evaluate and assist with imaging projects in order to ensure the most suitable technology is applied.
3. Continually study and discuss new developments in technology.
4. Partner with other public and private organizations to provide broad experience and resources to County departments.
5. Coordinate customer department contacts with vendors and/or the Purchasing Department.
6. Develop and present training as needed.
7. Provide an annual update to the Imaging Strategic Plan (see Appendix A).

Source: Hennepin County Information Services, *Imaging Strategic Plan*, 2 October 1992.

pens to a given document throughout a work process. Hennepin County has a 10-page "Imaging Requirements Interview Checklist" that examines in detail the following:

- Type of document, including quality, tone (black-and-white, grayscale or color) and quantity
- Retention period
- Copies made of the document
- How document is filed
- Whether document is ever disseminated outside of central site
- All aspects of document access
- User determination, including type of user, location, whether any people outside the county have access, and how often they retrieve a document
- Current computer usage and purpose
- Type of current database and its capability for image indexing
- Document archival process
- How documents are annotated, how the annotations are done and why
- Remote delivery of documents
- Workflow
- Printing/copying of documents
- Document security.[12]

A major purpose of the interviewing is to find possible deficiencies in the work process.

Analyzing requirements. With the current work process mapped out, the paper flow documented and the problems defined, the project team can begin to outline its requirements. Remember, the team should be trying to solve a business problem, not acquire a new technology. If the paper problem requires micrographics instead of imaging as a solution, then that's the route to take. Management shouldn't become enamored with the advantages of imaging and expect a vendor to make it work for a department.

During the strategic planning phase, goals should have been set. The team should know what the department's workflow looks like: how many documents arrive and are processed on a daily basis, how many people need to see them, how much is stored on-site and off-site and what the retention requirements are. This information is crucial because it allows the department to determine its load factor: the department's record storage requirements, the level (on-site, off-site) at which each record will be stored and for what length of time.[13]

If a department requires "real-time" access to its records—access identical to that on a typical data processing system—it will need the most expensive storage system: magnetic disk storage. If the department's access needs are not as time-sensitive—that is, if staff can wait 10 to 30 seconds to retrieve a document image—it may use less expensive on-line optical storage.

The slower response time of optical storage can, however, be mitigated in certain circumstances with a technique that is known as pre-fetching, which moves documents from optical disks to magnetic storage at a predetermined time—usually during off-work hours. Because magnetic hard drives are faster but more expensive than optical disks, they provide a short-term solution for speeding up the retrieval of selected document images. If that kind of retrieval time is unnecessary, then it's possible to store documents on either an offline optical storage system or a micrographics system, where costs will be lowest.

Of course, a department may need a combination of all three storage modes, since many documents go through a short, peak period of use, followed by a gradual decline in user demand. Such a pattern is typical in law enforcement agencies. Many police reports are used extensively during the first week in which a crime is reported; as time progresses, their use declines steadily. Imaging systems in many law enforcement agencies require a fairly large amount of magnetic storage space to handle the heavy use of document images among officers, investigators and detectives for newly reported crimes.

Assessing processing needs. Does the department need to process documents for the occasional ad hoc query or for hundreds or thousands of requests on a daily basis? Does the department want to replace filing cabinets with optical storage or is it planning to convert a paper-fed transaction system with image processing?

A few years ago, only a few options could satisfy the imaging needs of local governments, no matter what their requirements. Most were high-end, complex, expensive applications. Today, a number of mid-range and high-end systems are available, with lower prices and expanded capabilities and features.

Some government departments and agencies even have the luxury of choosing from function-specific software programs. Ready-to-go packages now exist for land records and registry of deeds offices, with built-in cash management systems. In law enforcement, vendors are offering imaging packages that link data from computer-aided dispatching systems with images stored in records management databases.

But while local governments now have more choices, they still face the problem of choosing the wrong system for the right need. An imaging project team must analyze such factors as indexing, processing, workflow, interface and database requirements, as well as other functional and organizational needs that can have an impact on the type of imaging software required.

Certain applications may require a great deal of customizing and applications development, which can only be supported by high-end systems. Then again, a mid-range imaging software program may meet all but one of a local government's requirements. If that one additional requirement means the difference between a medium-priced system and one that costs substantially more, it might be worthwhile to reassess that need and possibly drop it. However, if that requirement is driving much of the system's benefits, then it is wise to upgrade to the more expensive—but more valuable—software.

Centralized vs. decentralized systems. Linked to the issue of matching processing requirements with imaging needs is the issue of centralization versus decentralization. The project team must contend with organizational as well as technological questions in deciding whether to opt for a centralized or decentralized system.

In his book, *Image Processing: A Management Perspective,* author George Hall cites two factors for favoring a centralized imaging system: logical integrity and maximum relatability. "A centralized system," Hall says, "offers the best opportunity to ensure that data, information and records that are logically related remain in that state within the system."

Factors favoring decentralized imaging include security and greater freedom for local users. Hall points out that decentralized systems are less prone to catastrophic failure. Systems with multiple servers that handle specific aspects of imaging, such as scanning, the relational database and faxing, can, with some juggling, keep on running, should one of the servers fail. In a centralized system, however, everything shuts down when the central computer fails. Decentralized imaging systems also allow users to work without the imposition of a central data bureaucracy.

Cost and ease of integration are two other factors affecting the choice of centralized or decentralized systems. These last two points are, of course, crucial for local governments. Distributed systems, in theory, are less costly, since they depend on many inexpensive computers to function. Their costs can, however, escalate with network management demands.

Centralized imaging systems have higher hardware and, sometimes, software costs. Networking in these systems presents other, less costly, issues. For many government imaging applications, integration with existing databases is a must. Huge databases of public information stored in legacy mainframes need to be tied to document images in order to function more effectively. Yet integrating various databases by way of a consolidated index could be a monumental undertaking.[14]

Step 4: Preparing an Implementation Plan

After careful evaluation of needs and resources, a prospective local government imaging user must prepare an implementation plan that covers the imaging system's functional requirements. In this phase of the project, it is tempting to come to a quick decision on a preferred solution. However, it is still too early to think about product—first, the process must be finalized.

The following activities are key components of the implementation planning process:

- **Identify the major imaging applications based on the needs assessment.**
- **Identify an early winner.** Pick an application where the need for imaging and the payback potential are both high.
- **Identify any known limitations.** Be sure that imaging requirements don't limit your choices in hardware or software.

- **Identify back-file conversion needs.** Back-file conversion entails significant costs, so budget accordingly and take whatever steps you can to reduce the time and money involved. You may need to convert back files before your system can be up and running, but does every piece of paper need to be imaged? After purging your paper files, your storage needs may drop dramatically.
- **Identify links or enhancements to existing systems.** Designing an imaging system without adequate links to existing networks or databases is now considered short-sighted. If an existing database is too "old" to link with a new imaging system, examine possible ways to migrate the data to the new system.
- **Identify project benefits, costs and risks.** By comparing applications and assessing the experiences of organizations similar to yours, you can gather solid information on imaging costs and benefits. It is important to draw upon the experiences of other jurisdictions and identify the risks of implementation.
- **Identify project timeframes and milestones.** Plan adequately for procurement, negotiations, a possible pilot and final implementation, all of which require a substantial amount of time.

When the technical team presents a preliminary implementation plan to the policy team, it should suggest the total duration of the project and approximate annual costs. These recommendations will be based on very rough estimates, but can help gain top management and elected official support for expected funding.

If the scope of the imaging project is to be large—multi-departmental or enterprise-wide—it is necessary to establish a project management structure. Some organizations create a separate unit or department to manage the imaging project. This raises the question of who will control the imaging system. Should the data processing department manage the project, since imaging is a computer system? Or will the records management department control the system, because of its experience with records and documents? Control should not be the deciding factor, however, but responsibility based on the best fit. For example, technical aspects can be assigned to information service staff, while application functions can be assigned to individual departments.

Step 5: Beginning the Procurement Process

At this stage, the policy team reviews the preliminary implementation plan and recommendations made by the technical team and consultants or integrators (if the latter have become involved with the project by this point). The policy team's function is to clearly outline the expectations of management in the implementation stage. At the same time, the policy team can take appropriate steps to ensure that management supports the imaging system's long-term strategic goals.

By using specific guidelines, management can ensure that the system will become a shared resource. Organizations with no experience in data-sharing may find acceptable models of interoffice or interdepartmental cooperation in instances where offices or departments worked together to solve a cross-cutting problem.

At this point, management must make a preliminary decision on the level and duration of funding. The policy team also needs to build a conceptual design for the project. This can be done in-house or by a consultant or systems integrator. The conceptual design will help the team develop a well-working request for proposals (RFP).

Step 6: Preparing a Request for Proposals

The RFP is a local government's opportunity to enter into competitive negotiation with vendors. This is a time-consuming, complex process, where things can go wrong at any number of points.

Robert C. Cary, an imaging specialist with Pierce County, Wash., has managed the implementation of imaging on an enterprise basis across all departments in the county. Here are his basic points of advice for developing an RFP:

- **The RFP is a process that must be managed like any other.** It involves writing, communicating, amending, evaluating responses, benchmarking, negotiating and selecting a vendor.
- **The RFP process needs tools.** Word processing, project management, spreadsheet, graphics and communications software are the basic but necessary tools for pulling an RFP together.
- **Spend time on writing the RFP.** Understand what you want, organize your information coherently, keep it simple, write in clear English and be sure to build in an evaluation process.
- **Evaluate the proposals along the following parameters:** cost, functional excellence, technical excellence, vendor strength, references and benchmarks.
- **Negotiate with a win-win attitude.** View the vendor as someone to add to your team.[15]

For a detailed review of RFP preparation and announcement, including RFP types, content, format and terminology and evaluation tools, see Appendix C.

Step 7: Evaluating and Selecting Vendors

Bid evaluation begins by qualifying bids that meet a minimum score according to mandatory criteria. The goal at this point is to consider as many bids as possible and to invite responsive vendors to make presentations. In selecting finalists, it is important to rely on the needs assessment and the priorities set by the policy team.

Once the finalists have been selected, an in-depth review begins. Be sure to structure the review so that the team compares "apples to apples" as it considers vendor offers. Evaluation should follow a step-by-step process, so officials can refer to that process if any bid protests should occur.

At this point, benchmark testing should be conducted, and the first-choice vendor selected. The jurisdiction must prepare an evaluation package that summarizes by evaluation category the implications of the final vendors' proposals. This ensures agreement on the facts and guarantees that each system's potential trade-offs are clarified and compared with the policy team's priorities.

Step 8: Negotiating and Awarding a Contract
Hardware, software, support services, applications software development, consulting, training, price and terms of payment are all subject to review and negotiation. To achieve the maximum benefit from this phase of the project, it's important to have a negotiation plan, the contents of which can vary widely. However, several common guidelines are worth noting:

- **Have an objective.** Possible objectives might include reduction of item pricing, hardware or software loans, trial periods or evaluation period discounts, additional no-cost installation assistance, consulting, training, telephone support or financial gain from the local government's participation in software development.
- **Analyze risks and costs.** The client must share with the prospective vendor all risks and costs that are unacceptable to the jurisdiction.
- **Show commitment.** A list of all local government resources that will be used during the project implementation is important, as it will help to assure the vendor that the client's commitment to successful implementation will minimize the vendor's risks.

Once the jurisdiction has negotiated satisfactory terms, it can review and award the contract. The following aspects of imaging contract preparation deserve special attention:

- **Revision of the implementation plan.** The vendor or integrator can help a jurisdiction make necessary adjustments to the implementation plan by refining needs into specific tasks with clear and attainable objectives and by linking specific deliverables and acceptance test results to the contract.
- **Deliverables review.** A careful, formal review of the deliverables described in the proposed contract—including hardware, software and network services—must be performed.
- **Cost review.** Costs associated with the deliverables in the proposed contract must be reviewed. It is not advisable to proceed to the legal review until agreement on all costs is reached.
- **Legal review.** The process of legal review is subject to many local differences in purchasing regulations and practices. Ample (but not excessive) time must be allocated for this stage, especially if the contract contains provisions for dealing with data rights, copyright laws, cost of revenue-sharing or royalties and fees.

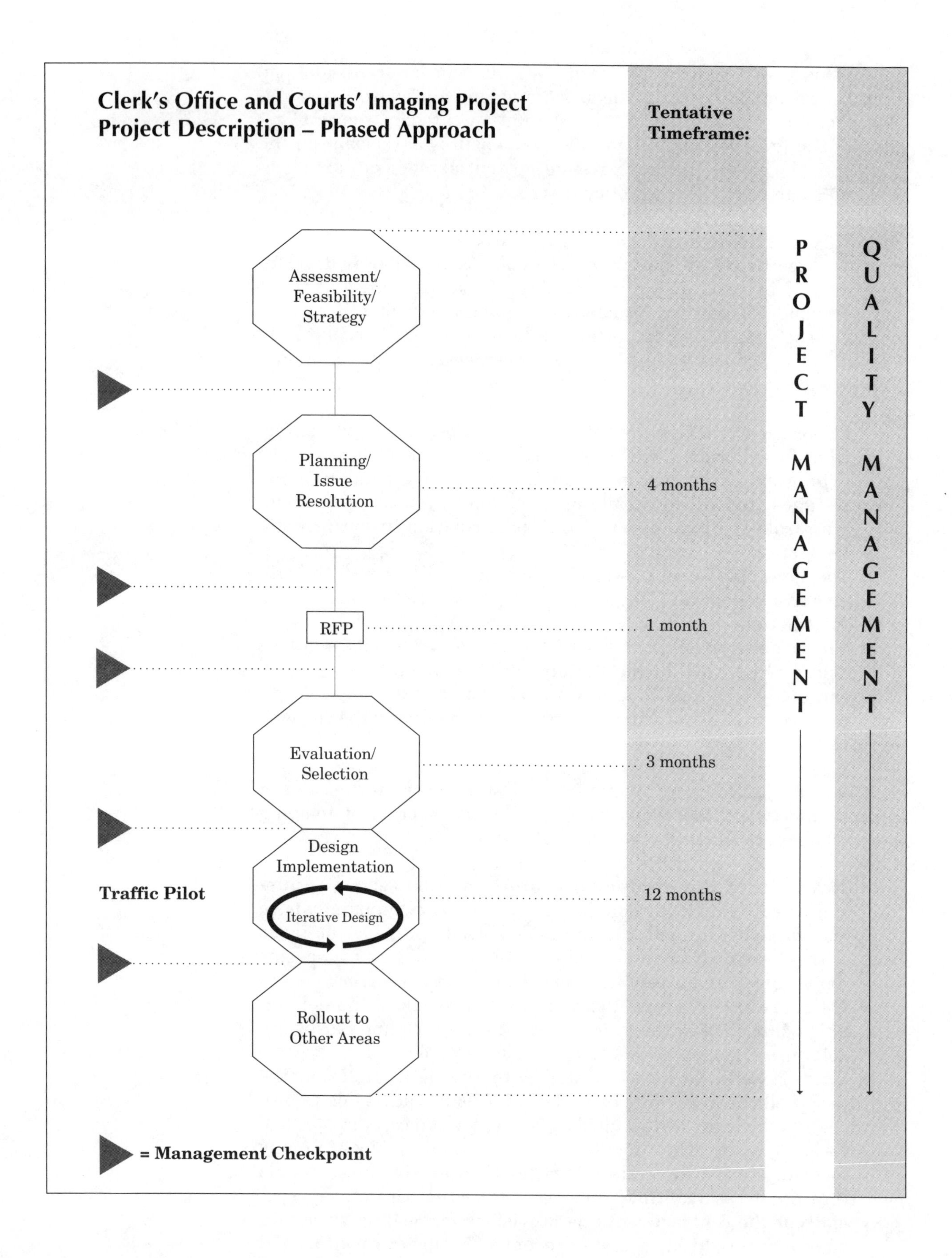

Source: Andersen Consulting

Installation

The Pilot

In the early days of imaging, when the technology was still emerging from its embryonic stage, many local governments began implementation by setting up a pilot to assure themselves that the technology did indeed work. Those early pilots were, in fact, prototypes to test or experiment with the technology.

Today, however, the purpose of pilots has changed. A pilot should help you determine how imaging best fits into your solution. As such, it should be an integral first phase of implementation, used to observe and study the impact of imaging before rolling a system out for the entire department or organization. For example, a pilot can be used to gauge the traffic imaging produces on a segment of computer network, testing the impact of the technology on network traffic without subjecting the entire network to the test.

Pilots should be conducted only when expectations for success and benefits are high and projected risks low.

Phasing

Phasing is considered very important in implementing an imaging system. In contrast to a pilot, phasing is the planned rollout of the entire imaging system. The goal in phasing is to carefully deploy the system for full-scale production in the department or enterprise. It allows an initial period when small errors can be caught and fixed before they become too large and expensive to correct.

Tom Carroll, project director for imaging systems at the state of Washington's Department of Labor and Industries, says phasing was key to the success of his department's implementation of a department-wide system. "Phasing doesn't let you fall into the trap of implementing a system based on bad data," he explained. For example, during the workflow analysis stage, employees told interviewers that they used a multi-page document required by the state, when, in fact, they only used the first page. This fact became apparent during the first phase of implementation, altering the department's document storage needs considerably.

Workgroup imaging systems rarely have more than one installation phase, while enterprise systems have considerably more. Within an enterprise system, each department-level installation involves several phases. All must be closely coordinated.[16] Phasing approaches include the following:

- **Building-block approach.** The most common option for phasing in an implementation—one complete unit is installed before any others are put into place.
- **Cross-section approach.** Starts with pilots in different functional areas of a departmental system, or in different departments of an enterprise system.
- **Smokestack approach.** Installs pilots simultaneously at different levels of an enterprise system.[17]

---

[1] Hennepin County Information Services, *Imaging Strategic Plan*, 2 October 1992, p. 11.

[2] Michelle Miller, "Document Management Projects: Managing the Organization Changes," *Workgroup Computing Report*, May 1994, pp. 2-3.

[3] John Martin, "Reengineering Government," *Governing*, March 1993, p. 27.

[4] Ibid.

[5] Scott Wallace, *Implementing Electronic Imaging: A Management Perspective* (Warwick, Mass.: Londahl & Wallace, 1993), p. 81.

[6] M.J. Richter, "A Guide to Planning for Imaging," *Governing*, April 1994, p. 76.

[7] Thomas E. Carroll, "Risk Mitigation Strategies for Successful Imaging System Implementation," 1994 AIIM Show and Conference, New York, April 19, 1994.

[8] *The Records and Retrieval Report*, April 1992.

[9] Wallace, p. 87-88.

[10] Ibid., p. 53.

[11] Ibid., p. 17.

[12] Hennepin County Information Services, *Imaging Requirements Interview Checklist,* 1 September 1992, pp. 1-10.

[13] George M. Hall, *Image Processing: A Management Perspective* (New York: McGraw-Hill, 1991), p. 91.

[14] Ibid.

[15] Robert C. Cary, "Developing a Request for Proposals for an Enterprisewide Image System," 1994 AIIM Show and Conference, New York, April 20, 1994.

[16] Hall, p. 127.

[17] Ibid., pp. 128-129.

Chapter 4

4 Technological Issues

Introduction

Imaging systems consist of hardware and software integrated to capture, store, display and disseminate images of documents, forms, graphics and photographs. The systems involve all the usual elements of today's information technology, such as input, output, data storage, database management, networking and client/server. They also use technology unique to imaging, including scanners, optical character recognition (OCR), optical jukeboxes, compression and workflow.

Like information technology in general, imaging has undergone rapid changes, especially in the areas of performance and cost. Mainframe and minicomputer imaging systems of the late 1980s have been replaced, to a large extent, by less costly personal computer systems that rely on local-area networks. Jukeboxes that contain optical disks work faster than ever; indexing has become easier, thanks to barcodes, OCR and intelligent character recognition (ICR); scanners scan faster; workflow software has become better at setting rules and procedures for routing document images through an organization; and the software that drives an imaging system has become easier to use.

Despite all of these advancements, imaging remains a complex, evolving technology. Large storage requirements, back-file conversion, image compression, indexing, networking capacity, workflow and high-resolution monitors are just some of the technological issues that must be tackled each time an imaging system is implemented.

As for standards, imaging still has a way to go. Some de facto standards for compression, operating systems and networking have been established, but gaps remain, especially in the areas of optical disk technology and integration between high-end and low-end systems. Open systems, where hardware and software become interchangeable, remain in most cases a goal, not a reality. Given all this, it's impossible to implement imaging and avoid upgrades and retooling at some future date. But while some might see too much risk in acquiring a technology that has yet to mature, a growing number of local governments believe the benefits far outweigh the risks and have implemented imaging.

To reduce mistakes, local governments need to fully understand the capabilities and limitations of each component in an imaging system. Imaging customers can choose from a large number of hardware and software products, each of which has strengths and weaknesses as well as advantages and disadvantages according to the needs of the system.

Input

Imaging begins with input: converting paper documents into bit-mapped images consisting of millions of light and dark pixels for storage and dissemination through computer systems. Input is the first of several critical operations in an imaging system where a wrong hardware decision can make a vast difference in the cost and performance of the overall system.

When a one-page, 400-word letter is keyed into a computer, it becomes computer-readable data in a code known as ASCII. The data occupies approximately three K of space in the computer's storage system. A scanner with a resolution capability of 200 dots per inch (d.p.i.) will convert the same one-page, 8.5 x 11-inch letter into a bit-mapped image consisting of over 4,000,000 pixels, each of which is equal to one bit of computer code, or roughly 500K of uncompressed data. The enormous data size of scanned images of documents (compared to the size of the ASCII version) is of crucial importance all the way down the imaging chain. However, the size of document images can be altered and controlled in several ways.

Scanners

A scanner is a piece of equipment capable of capturing images of documents at different resolutions. Low-resolution scanners—usually 200 to 300 d.p.i.—capture images with fewer pixels and smaller data capacities. However, the downside is less detail and quality in the image. High-resolution scanners—400 to 600 d.p.i.—capture images in fine detail, but the images they produce take up more storage space and take longer to transmit.

Since low-resolution images are harder to read, they can tire viewers, slow reading speeds and, ultimately, lower productivity—particularly in users who view images over long periods. Documents that contain lots of handwriting or finely detailed graphics compound the problem, ending up as poor-quality images if scanned at low resolution.

This is one major reason why it is so important to an-

Scanning workstations are used for quality control during the scanning process, allowing workers to view and, if necessary, reject images immediately after they are scanned.

alyze both document types and their use by workers when considering scanners, scanning resolutions and storage requirements. A jurisdiction can minimize storage needs by scanning documents at lower resolutions only if its imaging application involves highly readable documents and a work process with short periods of document viewing.

Different types of scanners not only capture document images with varying degrees of resolution, but also employ different techniques for capturing images. Some scanners are called flatbeds, with documents laid on top of a sheet of glass, much as in the common photocopier. Others are sheetfed, moving a paper document past a fixed scanning point to produce the image. Sheetfed scanners are better at handling large volumes of paper, but often cost more. Flatbed scanners (which usually have a document feeder added to automate scanning) cost less, but are not as fast as sheetfed models. For very minor scanning, typically of news articles or of small pieces of artwork, handheld scanners are capable of doing a respectable job.

Document Considerations

Document size and quantity are two other important factors to consider when choosing a scanner. Small documents, such as checks, and oversized ones, such as large drawings, require specialized scanners. Images that contain gray scale or color require scanners that can convert them into higher-density electronic images. High-volume applications require scanners with good mechanics, such as sheet feeders and two-sided scanning.

Sometimes images are scanned in at slightly crooked angles. Skewed images strain users' eyes, but more important, they can have a detrimental effect on OCR. High-end scanners can minimize this problem with features such as border detection and skew correction (known as "de-skewing"). These features drive up the cost of a scanner, but when weighed against the need for human operators to carry out the same functions, the investment might be worth it.

Compression

The data size of document images can be controlled through compression techniques as well as resolution adjustments. Virtually all imaging programs available today offer some kind of standard or proprietary method for compressing document images.

Without compression, storage systems would quickly fill up and the transmission of images over networks would be unacceptably slow. Developed by the fax industry in the mid-1970s, compression technology uses algorithms to convert a digital image of pixels into a mathematical code that can be stored much more compactly. A one-page letter that a scanner converts into 500K of digital data becomes about 50K, on a 10-to-1 reduction ratio, through the use of compression.

Standards

Different standards have been set for compressing black-and-white, grayscale and color images. CCITT (Consultative Committee on

International Telegraphy and Telephony) Group 4 is an international compression/decompression standard for black-and-white images. JPEG (Joint Photographic Experts Group) is a standard for compressing color photographs. MPEG (Moving Photographic Experts Group) is a standard for compressing video.

Besides reducing image file size, compression techniques—de facto, international and proprietary—enable images to be transmitted over networks and communication lines. TIFF (Tag Image File Format) is a de facto standard that is widely accepted among imaging users and is used to interchange digital image data, such as a document image, between different computing platforms. The federal government has established CALS (Computer-Aided Acquisition and Logistics Support) as an international standard format for the electronic interchange of document images, but that format has never quite caught on in the business community.

Proprietary compression techniques developed by various vendors sometimes do a better job than standards-based compressions, but governments should be aware that if the software that controls a proprietary compression becomes obsolete, or if the software vendor goes out of business, they run the risk of not being able to restore the image. Other concerns include matching document type with the correct compression technique and assessing the impact of compression on system performance. Certain compression standards can affect transmission speed, retrieval and viewing of documents. (See sidebar on "Software Volatility" on page 68.)

Indexing

Before a compressed image is stored in an imaging system, it must be indexed for retrieval purposes. Indexing involves linking descriptive information about an image with the image itself. It gives users a controlled vocabulary by which they can search for and retrieve an image or a file of images. The indexing information that is attached to the image is called the image header.

Central to the indexing process is identification of the items or fields of information that will be used to retrieve document images. For this, management must know something about the scanned documents. To work effectively, indexes need to be consistent and logical. The logic must enable staff to enter the minimum amount of information on a keyboard to retrieve the document image. Usually, the index logic for the imaging system will follow that of the pre-existing manual index system, if one was used.

Quality Control

Typically, scanned documents are indexed during or shortly after quality control. In quality control, images are visually inspected to ensure their readability. Government managers need to allocate enough time and manpower to ensure this job is done well. Distorted, high- or low-contrast or skewed document images will have an adverse impact on the indexing process (whether it's done manually or with

OCR), as well as on the legality of the document. A legal document that is unreadable because of poor scanning could create significant problems down the road.

Indexing itself needs quality control. One error in a multi-digit indexing number, such as a land title record number or a court docket number, will make it extremely difficult, if not impossible, to locate the document image.

Quality control and indexing are labor-intensive tasks that are not usually required in manual records management systems. Labor costs for indexing and quality control are unavoidable in a system designed to automate multiple manual tasks. In fact, a number of government users of imaging have remarked that the addition of work for indexing and quality control usually wipes out any savings the system has generated through the elimination of clerical filing.

To reduce the labor associated with these two tasks, vendors have turned to technology. High-end scanners offer software that can automatically adjust documents skewed in the scanning process, flip documents scanned upside down and adjust contrast to improve the readability of faint images.

Optical Character Recognition (OCR)

To facilitate the indexing process or replace it entirely, vendors have turned to OCR technology. OCR is software that can convert the letters and numbers that appear on a bit-mapped image into computer-readable text. As a tool, OCR can automate the task of entering data into key fields for document indexing by reading names, addresses, social security numbers and other pertinent information directly from document images. At its best, OCR eliminates indexing labor, leaving only indexing quality control to human operators.

While OCR itself has become faster and more accurate in recent years, its function as a tool for automatic indexing relies heavily upon the quality and uniformity of the scanned documents. Since indexes require consistency and logic in order to work, an index with misspelled words or incorrect numerals is useless. Second-generation photocopies and documents printed on colored paper or containing boxes and lines are usually poor candidates for OCR. Inferior copy quality, color backgrounds and borders lower the chances that characters will be read correctly.

Beware of vendor claims that OCR can achieve a 90 percent recognition rate or better, implying that the verification workload will be minimal. A 90 percent recognition rate means 10 errors for every 100 characters read, making verification a laborious task. If accuracy is an issue, error rates shouldn't fall below 99 percent. Otherwise, a human operator should key in the index fields.

Another concern with OCR is its ability to locate index information on a document image. If that information is not in the same place on each document, then OCR cannot be used very readily for indexing. The problem can be solved if certain key fields of information appear in the same spot on each document. If an agency has control over the documents it scans, it can redesign them so that social security numbers, phone numbers, dates, addresses and other index-related items always appear in the same position. Proper redesign will enable more rapid and accurate readings. However, forms design is a complicated issue that needs to be addressed well before procurement or implementation begins.

Barcodes

For departments that can't control the formats of documents they receive and scan, one alternative is to use barcodes for indexing. Barcodes are highly accurate—99 percent on average—and can create an entire index for a document. Typically, the barcode technique is used for indexing batches of similar documents. A sheet of paper with a preprinted barcode is placed in front of a stack of documents, which is then scanned. The imaging system reads the barcode—usually a control number—then automatically links the number to each document for routing and retrieval purposes. Some systems print barcode labels, which can be affixed to documents before scanning. The labels are generated from a central database, creating a direct link between the documents and the database.

Text Retrieval

Another indexing alternative uses OCR to create a searchable text file from a document image. Instead of relying on a controlled, consistent index database, the system uses OCR and text retrieval software to create an index of all words in the document. Useful for research involving reports, council minutes, manuals or other material, OCR creates an unedited ASCII file of the document, which can be searched by virtually any keyword. When the keyword is entered into the imaging system, it provides the user with a list of hits. When an item is selected, the system retrieves the actual document for review.

ASCII files for text retrieval are usually stored on a computer's magnetic hard drive, where they can be retrieved quickly and edited, if necessary. While storage requirements for ASCII text are modest compared to those for document images, users of text retrieval should still plan their magnetic storage needs carefully.

Intelligent Character Recognition (ICR)

ICR is another indexing tool with potential in certain local government applications. Like OCR, ICR can be used to create indexes from a variety of government forms. Unlike OCR, ICR is designed to read handwriting, making it especially well-suited to reading forms typically filled out by the public, such as those for property, taxes, birth certificates, permits, licenses and vehicle customer surveys.

ICR does have limitations, however. It can only read handprint, not cursive handwriting. It works best in high-volume applications (those that process 500 forms or more per day), where input can be centralized, handprinting is controlled through the use of boxes and data manipulation is minimized.

Storage

The success of imaging is due in large part to the invention of the optical disk as a storage medium. Magnetic storage, which has been around much longer, remains too expensive a storage medium for the large quantities of document images in a typical local government imaging system. Optical disks, however, provide a very economical way to store and retrieve enormous quantities of images. Their single major flaw is their slow response time when compared to magnetic storage media.

Types of Optical Storage

Three optical disk types exist: CD-ROM (Compact Disc, Read-Only Memory), rewriteable and WORM (Write Once/Read Many).

CD-ROM. CD-ROM is used essentially for electronic publishing. Government agencies are using CD-ROM to distribute everything from county statistics and administration manuals to the latest in government regulations. The disc is a unique size, 4.75 inches; like its audio namesake, it is produced in quantities from a master for distribution to multiple sites. Because of the special equipment involved, most CD-ROM masters and copies are created by firms that specialize in their production. CD-ROM is incompatible with hardware that runs rewriteable or WORM disks.

Rewriteable disks. Rewriteable disks vary in size, and only work with specially designed hardware. This optical disk is ideal for use with material that needs frequent updating. There are currently two incompatible rewriteable techniques: magneto-optical and phase change. Magneto-optical systems combine properties of magnetic and optical technologies. Their one major drawback is increased recording time, since two passes across the disk are required to record data. Phase change provides a direct data overwrite capability, making recording time faster. But the technology currently lacks standards.

WORM disks. First introduced in the 1980s, WORM optical disks remain the most popular choice for imaging systems. Unlike CD-ROM, which requires costly special recording devices to store information, WORM disks are a relatively inexpensive, ready-to-use recording medium that the user can plug into a drive (much like floppy disks) and use to record data. WORM gives users control over what they wish to store on the disk—and when they want to store it—and keeps the data in a permanent, unchangeable state. WORM disks currently come in three sizes: 5.25 inches, 12 inches and 14 inches. Their storage capacity varies greatly, depending on the type of image stored, its resolution and the compression ratio used. If an agency is storing the typical 8.5 x 11-inch, black-and-white document scanned at 200 d.p.i., capacity can vary from 20,000 pages on a two-sided 5.25-inch disk to as much as 330,000 pages on a two-sided 14-inch disk.

The platter drive shown accommodates large, 12-inch optical disk platters, which can store as many as 330,000 document images and are well suited for high-volume, fast-response applications.

Disk Size

Twelve- and 14-inch disks are often used in high-volume transaction applications, where quick access is essential. Unfortunately, these large disks and their drives have no standards. As a result, a 12-inch optical platter from Vendor A will not run on Vendor B's 12-inch optical drive. The 5.25-inch platters are more standard (they can be swapped between different types of drives) and are cheaper than the larger disks. However, they are slower (fewer documents are stored on each disk; therefore, the jukebox must move more disks in and out of the drives for each retrieval). Be sure to carefully assess the application before choosing an optical disk size.

Optical Jukebox

Housed in a protective plastic cover, the optical disk is inserted into a special drive that both "burns in" an image (using laser technology) and retrieves it when requested by the user. Since most imaging applications involve dozens of disks, the drive is housed in a special jukebox, which has become the standard hardware for storing and playing many optical disks. The jukebox is a mechanical device, the size of a refrigerator, with a moveable gripper that pulls disks off shelves and inserts them into drives as users request specific images.

Optical jukeboxes contain stacks of optical disks, which are pulled from trays by a mechanical arm and mounted in a disk drive. Jukeboxes come in a range of sizes.

The jukebox provides mass storage capabilities in a small amount of space and at a relatively low price. Some jukeboxes hold as much

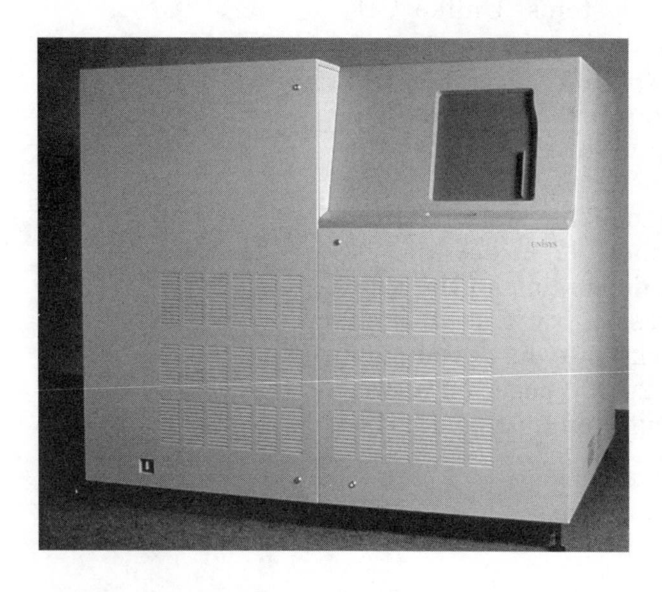

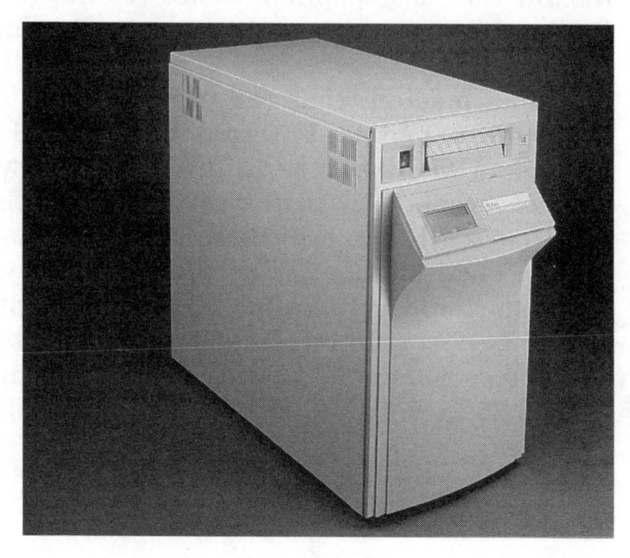

as a terabyte (one trillion bits) of data. Despite its capacity, compactness and economy, the average optical jukebox is demonstrably slower than a comparable magnetic storage device. Magnetic storage on disks is today's typical medium for data storage on personal computers.

Response Time

Slow response time is attributable to the way a jukebox finds the correct optical disk and mounts it in the drive, as well as to the way the drive locates and reads a document image stored on the disk. Response time can range from under 10 seconds to more than 30 seconds, and can be even longer during peak workload situations. When dozens of workers request document images almost simultaneously, the delays can escalate, since the jukebox is only able to load one disk into one drive at a time.

That's why many consultants tell buyers of imaging systems to make sure the application is designed to handle peak loads, to ask: What is the maximum number of users requesting documents at a given time, and how many documents do they request?

Micrographics

While the longevity of optical disk technology is debatable, microfilm, the alternative storage medium for paper documents, remains a stable, well-understood solution. For starters, the life of microfilm can exceed 100 years. Other big benefits of microfilm include its low cost and its admissibility in a court of law.

For local governments that have mandated the use of microfilm as a method of document backup, but still want to take advantage of imaging, there is hardware that can simultaneously convert paper documents into both microfilm and images for optical storage.

Microfilm is also used to record and archive electronic document images. The technology for this conversion is computer-output microfilm (COM), and it is used primarily to move little-used documents from faster but more costly computer storage to slower but less expensive microfilm. The conversion is performed by a device called a computer-output microfilmer. It records machine-readable data (such as ASCII text or computer images) directly to microfilm or microfiche without creating an intermediate paper copy. The advantages of COM include:

- The ability to distribute large quantities of printed images to users who don't have access to a department's imaging system, but have available low-cost microfilm viewers.
- Because COM contains human-readable information, it isn't dependent on hardware and software that might, at some future time, be discontinued, making the data inaccessible.
- COM can be more stable than the optical and magnetic media that store digitized images.

Microimagers transfer microfilm document images onto optical disks. They also can capture data from microfilm by reading the text with optical character recognition software.

Pre-fetching is one solution to slow jukebox response. If a department can predict its workload for each day, it can use software that automatically searches for and retrieves document images during off-peak hours. Pre-fetching moves desired document images from optical disks and stores them on magnetic disk drives at each imaging workstation. When employees arrive the following day, they can go right to work without waiting to fetch images from the jukebox.

In many law enforcement imaging applications, scanned police reports are kept entirely on magnetic storage during the first week or so, when demand for their use is likely to be highest. Once that period has elapsed, the images are automatically transferred to optical disks for permanent storage. This approach usually calls for large storage capacity on magnetic hard drives.

Under normal office conditions, the optical disk is fairly sturdy. The life of data stored on the disks is at least 10 years, and more likely to average 30 years under prescribed conditions of humidity and temperature. Government agencies must contend not with the question of how long the disks will last, but with this one: Will the technology that reads the data on an optical disk today still exist 30, 40 or 50 years hence? To ensure that the storage media keep pace with imaging technology over time, every government should establish clearly defined rules of retention. (See Chapter 2.)

Workstations and Servers

When it comes to viewing and working with the images and databases related to an imaging application, users can choose from a wide range of hardware, thanks to standardization and low costs. However, a few necessary features should be mentioned.

Workstations

Most vendors will recommend that users have a personal computer with an 80486 Intel microprocessor running the MS-DOS operating system with the graphical interface Windows. For heavy users of imaging, the monitor should be high-resolution and capable of displaying two pages of images, or an entire image and a database. Other monitor features to consider include the dot pitch (the smaller the better) and the refresh rate (the higher the better).

Servers

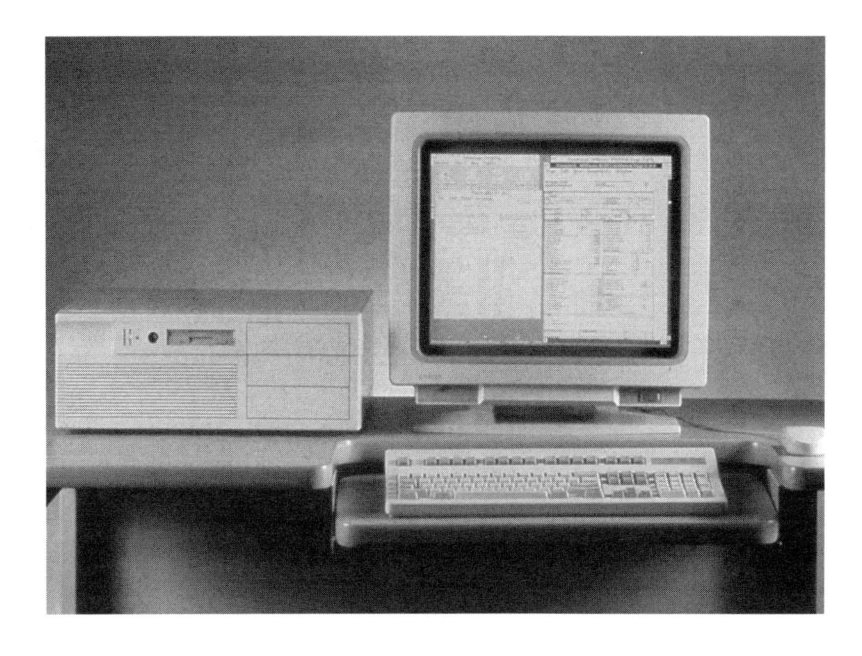

A server is a computer that stores data and manages the software for the imaging system or specific functions of the system, such as image management, database management, scanning, printing, faxing and OCR. The server can be a souped-up version of the industry-standard personal computer, using an Intel microprocessor, or it can use a different kind of microprocessor, known as RISC. While Intel servers usually run the operating system known as MS-DOS, both Intel and RISC servers also use an operating system known as UNIX. The type of server and operating system used depends on the kind of computing platform that is chosen for the imaging system, as well as the type of imaging application. UNIX servers are often used to link an imaging system with an existing mainframe database. Servers with DOS are often found in networks of personal computers.

For the best response time and viewing, users need a personal computer with a fast microprocessor and a high-resolution monitor capable of displaying two pages of images.

Printing and Faxing

Printing

At some point in the imaging cycle, users will need to print a copy of the document image. With the exception of the smallest applications, all government imaging systems will require both a laser printer and a print server. The laser printer is necessary for reproducing the graphical detail of stored documents (remember, the images are not text, but black-and-white pixel images of documents, including text and graphics).

Print servers manage the laser printing of documents on a network. They take care of the many requests from users for print copies of images and make sure the image server doesn't get bogged down with printing demands. Unfortunately, even with print servers, the process of printing document images can be slow. Occasionally, a user requests the printing of numerous documents. Print servers will download the images from the jukebox and queue them up for printing. Printing doesn't begin, however, until all the images are "spooled" or gathered on the print server. For large document files, this can take several minutes or more, delaying fulfillment of other printing requests.

Short of limiting printing requests to just a few pages, there's not a lot that can be done to eliminate this familiar bottleneck. Some vendors, however, have introduced software that can enhance both printing speeds and the spooling of document images. An imaging system can also support batch printing at off-peak hours and the printing of documents without prior display of their images on a monitor. For government departments that have volume printing requirements for their imaging systems, a networked printer or dual-path copier might be the answer (see sidebar below).

Faxing

The explosive spread of fax machines throughout the U.S. has given imaging an added tool for expanding its functions and capabilities. Limitations in current networks and telecommunications make it impractical to transmit a document image from an imaging system to another computer at a remote site. Not only is transmission speed a problem, but it is unlikely that the remote computer will be capable of receiving and displaying the image on its monitor. (The average desktop computer lacks the high-resolution monitor needed to display images.) It is, however, likely that the remote site has a fax machine, which is quite capable of receiving a document image in reasonable time and printing it.

Many government imaging applications include a fax service that disseminates document images to remote locations. Police departments use fax machines to transmit images of police reports to district attorney's offices and accident reports to insurance companies. Deed and land record offices use fax machines to transmit copies of titles and other land documents to law firms and title companies. If a department faxes images regularly, a dedicated fax server should be

Dual-Path Printing and Copying

Copiers and printers have traditionally performed separate functions. Printers were in individual offices and met low-volume personal printing needs. Some high-volume printers were linked to a network and served several users. Copiers offered finishing features, such as stapling and two-sided copying, that made duplication and finishing more productive. Dual-path copying and printing merges these previously separate functions.

Digital copiers permit networking. These networked copiers become multi-user printers with finishing capabilities. Sometimes referred to as multifunctional devices, these units offer desktop access to copiers from traditional personal computers, thereby eliminating unproductive steps and purchases. Previously separate printer and copier purchases can be combined. Previous "print in the office" and "walk the paper to the copier" functions can be combined, improving productivity and efficiency.

Source: Eastman Kodak Company

added to handle the computing demands associated with moving images from the jukebox to the fax machine, relieving the burden on other imaging and network services.

Fax machines and fax modems can also be used to receive document images for storage in an imaging system. A fax modem acts like a regular fax, but with one important difference: when it receives an image, it feeds that image directly into a computer instead of printing a paper copy. Some colleges and universities have used fax modems to enter student transcripts directly into an imaging system. Faxed images suffer in quality, however, because of the low resolution (200 d.p.i.) at which they are transmitted. This issue should be addressed before a department with an imaging system decides to rely on faxed images for its incoming documents.

Imaging Software

Despite the unique hardware requirements of imaging, software is the engine that makes everything move. In fact, as hardware costs continue to drop and networks of personal computers become ubiquitous tools in offices everywhere, software is becoming the central component of today's imaging systems.

Software manages the imaging system. It allows the different components to work together, furnishes the logic for locating or performing tasks and provides the interface between the system and its users. The typical capabilities of imaging software include scanning, indexing, storing, displaying and printing images.

High-End Software

Early imaging applications were based on the turnkey approach, with the software running on proprietary hardware. Because of the high cost at the time, imaging software was designed to run only large applications, where it was thought the investment could be justified. These systems were designed to manage large image processing tasks and had the horsepower to handle all the necessary hardware interfaces and device drivers, as well as database management and workflow needs. Today, the high-end software systems remain, but they now are designed to be open, meaning they can operate in conjunction with the variety of hardware and software most often found in today's offices.

LAN-Based Software

Meanwhile, the rapid increase in installed local-area networks (LANs) has fueled

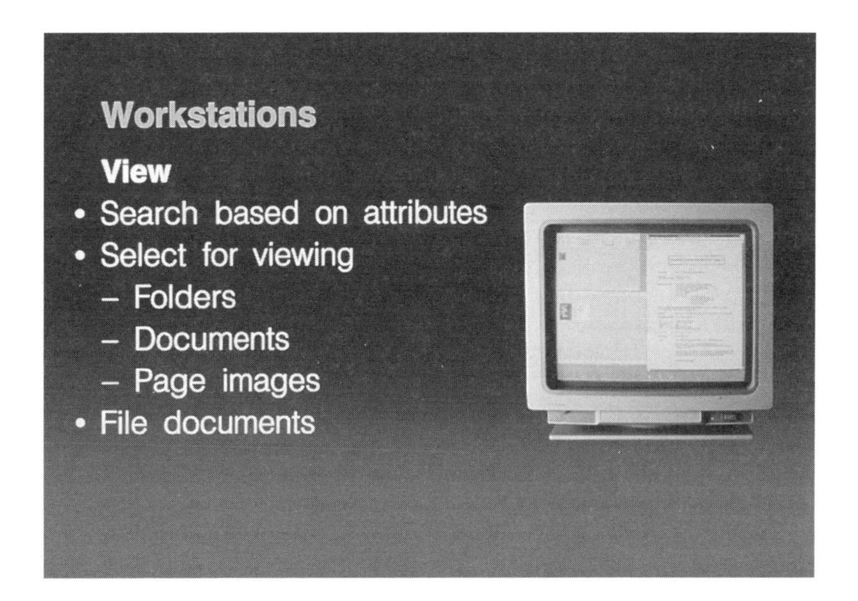

Imaging users work from personal computers, often called view workstations, that can retrieve and display images from optical disks, as well as perform other tasks, such as index searches, database management and word processing.

the market for LAN-based imaging systems. Both the early, high-end and a whole new breed of medium-range and low-end imaging software products are now available for a wide range of applications. Many new LAN-based imaging systems are designed to operate with the popular graphical user interface known as Microsoft Windows and the network operating system Novell NetWare. Rather than incorporate their own database management systems and workflow, LAN-based imaging programs are designed to work with the most popular models on the market. These LAN-based programs have incorporated some or all of the aspects of client/server computing.

Software Volatility

Today's imaging systems often consist of many pieces of software: the application that stores, retrieves and distributes the images; the relational database that contains the index or searchable data for finding images; and the underlying operating system. All of these software programs are constantly upgraded to boost their performance, features and functionality.

Unfortunately, upgrades can get out of sync. When an operating system, such as UNIX, gets an upgrade, an imaging software vendor may be slow to provide a corresponding upgrade that keeps everything compatible. That situation can create havoc for an imaging system. To prevent such problems, some local governments purposely have chosen well-established vendors. But even some well-known vendors have been known to discontinue certain lines of application software, such as imaging.

To protect a system against disruptive change, a jurisdiction can work with a systems integrator who assumes the challenge of keeping software components in sync. A local government can also ask the vendor who develops the imaging application to provide a demonstrable migration path away from its software, drawing up a contract that requires the vendor to show exactly how to find an image or data and transfer it out of the system for storage elsewhere. It is especially wise for a city or county to do so when working with small vendors, which are more likely than larger firms to go out of business.

The local government can also incorporate into any document retention policy a stipulation that a contract with an imaging system vendor or systems integrator will detail who is responsible for keeping the software components of the system up-to-date and how, ensuring that document access is never in jeopardy. (See also retention issues in Chapter 2.)

Another approach is to maintain staff skills so that knowledge about the system remains with the department, independent of the vendor. Finally, local governments can stay away from "bleeding-edge" technology. Don't go with the latest relational database or an entirely new kind of imaging software. Stick with the tried and true.

Image-Enabled Software

Image-enabled software is a recent addition to the imaging software market. These low-cost programs add imaging capabilities to other software products, such as database management systems, spreadsheets, electronic mail systems and workgroup programs. The practicality of image-enabled software is under debate in the software community, mainly because these products lack the ability to link up with larger imaging systems. Given the rapid development of the technology, however, the situation is sure to change. (See Chapter 5.)

Networks

Networking and connectivity are the fastest-growing areas of information technology in government today. The economies of scale and cost savings achieved through the sharing of both information and hardware devices, such as printers, among offices filled with personal computers, have driven the rapid growth of networks.

With the growth of networks, the demand for imaging on LANs has surged. Vendors have responded to the sudden interest. As a result, most imaging systems are designed to operate across the most common network types and on interconnected LANs, known as wide-area networks. With its large data files, imaging does, however, present unique challenges to networks.

Local-Area Networks (LANs)

Today's typical LAN operates over wire cables using one of two access methods: Ethernet or Token Ring. For the purpose of imaging, the main differences between the two are their layout or topology, cost and bandwidth. Ethernet uses an inline or bus topology and costs less, while Token Ring uses an easy-to-configure star arrangement and has more bandwidth.

If a LAN is already installed in an office or department, the network often must be reconfigured to support imaging. Some have compared putting imaging on a network to trying to push a basketball through a garden hose. Networks were designed to transmit small files from word processing, spreadsheets and database management programs between computers in small bursts of data. They were not designed to handle the much larger document image files that an imaging system creates (remember that one page of ASCII text contains 3K of data, but the same page as a compressed image contains 50K of data). If half the employees on a LAN were

A computer network allows users to share not just images and data, but also the components of an imaging system, such as scanners, jukeboxes, fax/modems and printers.

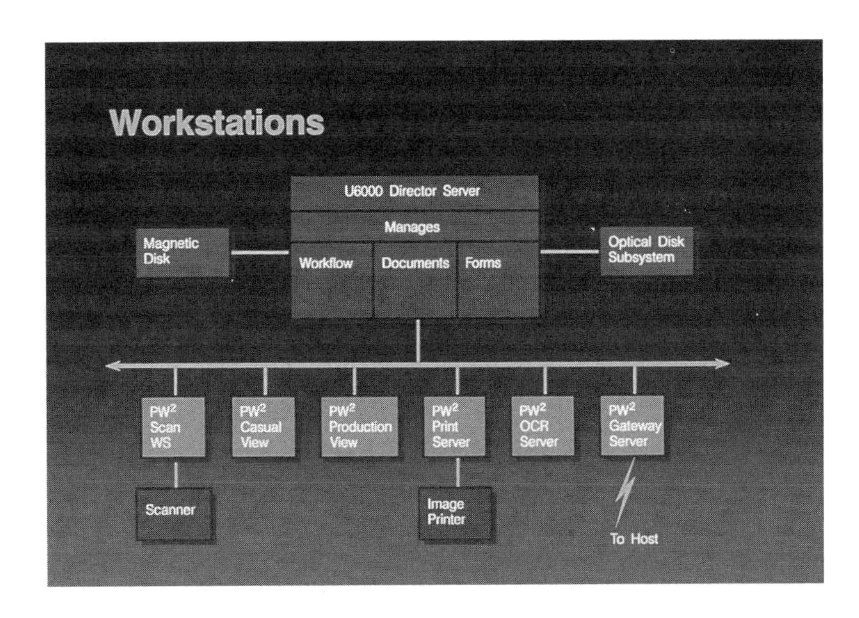

to suddenly use imaging, the network would probably grind to a halt at certain points in the day.

To prevent such bottlenecks, LANs need to be redesigned so that the addition of an imaging system doesn't disrupt other computing traffic. The most popular approach is to segment the LAN into parallel sections, much like adding car pool lanes to a highway to keep faster-moving traffic from interfering with slower-moving vehicles. Segmenting is done with devices called hubs and routers.

Other ways to control imaging systems on LANs include configuring the system so that certain tasks, such as scanning and printing, don't interfere with the running of business applications; adding separate servers for individual imaging functions, such as storage and retrieval, database management, OCR, faxing, scanning and printing; and using imaging systems that are specifically designed to take advantage of the features of the network operating system. Certain imaging systems, for example, are closely linked with Novell's popular NetWare operating system for LANs. Others are designed to work well with UNIX.

If a department is building its first LAN with the introduction of imaging or foresees significant expansion in the use of imaging throughout an existing LAN, it is worth considering some of the faster network access methods now available. These include Fast Ethernet (a much improved version of Ethernet) and FDDI, a form of fiber-optic network. Both significantly increase the traffic capacity of the LAN, but at a stiff price.

Client/Server Systems

Client/server systems take advantage of LANs to deliver computer applications at lower costs and with greater versatility than was ever possible with host-based systems, such as mainframes. Client/server goes beyond the mere sharing of files that takes place on most LANs today and, instead, distributes the workload.

Client/server computing allows users to perform individual tasks on their desktop computers and access information that resides on other computers in a networked system. Users are able to obtain and manipulate data faster and make better-informed decisions.

In a typical client/server model, the "client," the personal computer or workstation on the user's desktop, is connected via a network to the "server," which contains the departmental data the user requests through his or her desktop computer. The client includes the software that presents data to the user in a logical and easy-to-use format. It also runs popular personal computer-based applications for such tasks as word processing and spreadsheet calculations.

While definitions of client/server vary, the model's true measure shows up in the benefits it can deliver. For instance, when a user on a typical LAN imaging system wants to print a document image, the computer momentarily blocks other actions while the application converts the image into a format the printer can use. A client/server imaging system would immediately pass this function to a print server on the network, freeing the desktop computer for other work.[1]

Other advantages of client/server computing include incremental growth (users can begin with just one client workstation), faster application development and greater productivity gains. Applications are also easier to maintain in a client/server environment. If one component of the client/server system needs maintenance, the problem is isolated to that segment. If a mainframe is in trouble, however, the whole system is in trouble.

Wide-Area Networks (WANs)

When an imaging application calls for sharing images among several LANs in a local government department, the department must address the issue of connectivity as well as that of transmission speeds. Several approaches exist for creating a collection of LANs, known as a wide-area network (WAN). What users of imaging systems must consider is the compromise in transmission speed they are willing to accept in order to have an affordable wide-area network. The slowest link between LANs is the dial-up modem; the fastest is fiber. In between are a number of options with varying transmission speeds.

The costs of these faster transmission methods can escalate rapidly. Some options, such as frame relay, which provides high-speed transmission on demand, give local governments an alternative to expensive dedicated high-speed telecommunications. Other options, such as asynchronous transfer mode (ATM), are still several years shy of technological stability.

Integration: Links to Existing Databases

Crucial to the success of local government imaging is integration with existing applications. Most so-called "mission-critical" databases in local government remain on mainframes and minicomputers, often called host computers. Database contents range from voting records, law enforcement files, property assessment data and birth and death information to payroll, finance, administrative and utility records. Information in all of these databases is almost always related to document images, and all major imaging programs have tools that can link those document images directly to related databases.

Depending on the tools, integration can range from the simple to the complex. In a simple format, known as terminal emulation, the image document and the host database can be viewed simultaneously, but the two windows of data are not integrated in any manner. In a more complex integration, information can be transferred from the host database to the imaging application database and vice versa. For example, an employee number from a host payroll system could be passed to a query screen in the imaging application.

The impact of integration can be tremendous. Local governments can keep their legacy computers and databases and still use imaging as an information management strategy. For government departments with large databases and complex applications residing on existing computers, this is of critical importance.

A licensing or permit bureau, for example, stores many thousands—even millions—of names, addresses and license numbers in a database on a mainframe computer, and keeps comparable numbers of paper documents, application forms and correspondence in filing cabinets. By converting the paper documents into images and linking them with the host database, the bureau offers its employees a single point of access to necessary information and enhances their productivity.

Workflow

Imaging has allowed staff to break out of the serial work cycle, where documents move through an organization via "in" and "out" baskets, one employee at a time, and into parallel work processing, where more than one employee can act on the same piece of information simultaneously. By automating the routing of images between employees and controlling the work process, workflow software allows imaging users to take advantage of parallel processing.

According to Grace Kaplan, an IBM consultant on work management, workflow does more than automate the routing of work. "Workflow software automates and controls the entire scope of work, including how work is defined, designed, prioritized and distributed between people in a workgroup in separate departments or across an entire organization."[2]

Workflow software of varying capabilities is available both as part of an imaging system and as a third-party tool. The latest versions of workflow are highly intuitive, object-oriented programs that allow users to assign tasks to employees, who in turn are prompted or instructed by the workflow software to perform certain processes. Workflow software is not dependent on imaging. The technology can be used to streamline and galvanize any work process.

Vendors of workflow software say that their tools help an organization focus on the work, while workflow takes care of the process. That can trigger better services and higher productivity. But workflow, like imaging, is not plug-and-play software. It forces organizations to reexamine their work processes and change the ways their employees ordinarily perform tasks.

Workflow and imaging could, for example, help a claims department in which any claim of over $5,000 must receive supervisor approval. Workflow software can look at the value of the claim, and, if it exceeds the limit, automatically route the claim's image to the supervisor for the necessary action.

At an unemployment office where 20 percent of all claims must be audited, a workflow tool could automatically route the right percentage of claims to investigators on a regular basis.

Ergonomics

While the use of computers is fairly pervasive in local government today, some work environments use computers extensively for the first time when they implement imaging systems. The change from working with paper eight hours a day to viewing documents on a computer screen for the same period of time can have physical drawbacks.

Workflow and Police Records

Applications of intelligent workflow and imaging are still evolving in state and local government, but the city of Stockton, Calif., police department is a trendsetter. In 1993, the department introduced imaging to reduce document photocopying and speed the dissemination of time-critical police reports to investigators.

One departmental priority was to automate the flow of document images from records management to the appropriate investigative division within the department. An image management and workflow tool from Data Management Design, Inc., now prompts record clerks to answer a series of questions about the reports they have scanned: whether or not, for example, a report involves a juvenile, a stolen vehicle or domestic violence.

Based on the answers, the workflow software determines where images of the report will go, then automatically routes them to an electronic "in" basket in the appropriate division. Reports that must go to another city department or to an outside agency are routed to their recipients via a fax gateway, which contains key phone numbers for automatic dial-up and faxing. Coupled with workflow, imaging has enabled the Stockton police department to absorb an increase in police reports without jeopardizing the flow of information or increasing staff.[3]

The longer employees work with computers, the more likely they are to suffer from eyestrain, headaches, backaches and double vision. In addition to being a health concern, these physical ailments can reduce productivity.

Eyestrain is the most common computer-related health problem. It can have a variety of causes, including improper overhead lighting, glare from windows or lights, poor workstation arrangement and poor screen design (contrast, resolution, flicker rate, etc.). Some of these problems can be solved by rearranging furniture, putting blinds on windows or changing the wattage on light bulbs, but choosing a well-designed video monitor can make a big difference. According to Dr. James E. Sheedy, an expert on visual performance, the following are desirable attributes of a computer monitor:

- High resolution (results in fewer adverse effects)
- Gray scale (improves the visibility of displayed text)
- Dark characters on a light background (improves work performance)
- High refresh rate (screens redraw or refresh themselves many times per second—the faster the better for the eyes)
- Small dot pitch for color monitors (color monitors use grids to display colors; the smaller the grids—or dot pitch the better defined the screen image).[4]

Security/Disaster Recovery

Three major security problems associated with all forms of data processing are protection from physical damage, reduction of input errors and control of unauthorized access. Imaging systems create three additional security issues: the length of time over which records are kept, the sheer mass of those records and the records' accessibility to a much larger number of personnel than was possible before imaging.[5]

Physical Damage

As for protection from physical damage, optical disks are, fortunately, more stable than magnetic media. Head crashes, the bane of hard disk drives and a chief cause of physical damage to data, are all but impossible with optical disks, because the light source is much further from an optical disk's surface than it is from a magnetic head. Unfortunately, indexes for imaging systems are usually kept on magnetic media, where they can be updated. This is a weak point in image processing, so backup should be routine to avoid data loss.[6]

Input Errors

Input errors in imaging take two forms: 1) poor scans of documents that go undetected until after the original documents have been destroyed, leaving only unreadable document images, and 2) numerical or other errors in or long-term data decay to the index that makes it difficult, if not impossible, to locate document images. Document images must be checked for scan quality, and indexes perused for data accuracy and, periodically, for deteriorating magnetism on random bits of data.[7] Manual verification is the only way to do this. How much verification is necessary depends on the extent to which the documents will be used in court proceedings or are a part of permanent public records, such as deeds, titles, birth certificates, etc.[8]

Unauthorized Access

Unauthorized access can be controlled through passwords of varying complexity. Most network operating systems, as well as imaging systems, offer a range of password and viewing control techniques. For government departments, these not only prevent unauthorized attempts by either employees or hackers to access documents, but also allow public access to some document images while protecting other sensitive documents and indexes from abuse.

For many organizations, imaging heightens the importance of security. Imaging not only improves access to documents, it can expand the number of documents to which workers and the public will have access and the length of time for which they will be accessible. Some documents, undoubtedly, are more sensitive than others. Government agencies need to assess the content of documents that will be available via imaging, set procedures to control access to sensitive documents and review the duration of those documents' availability.

---

[1] Richard Barrett, "Client/Server Imaging Systems: A Term That's Lost Its Meaning," *Imaging World*, 23 May 1994, p. 8.

[2] Tod Newcombe, "Workflow: The Force Within Imaging," *Government Technology*, April 1994, p. 52.

[3] Ibid., p. 53.

[4] James E. Sheedy, "Visual Ergonomics: What's Important," in *1994 AIIM Conference Handbook* (Silver Spring, Md.: AIIM, 1994), p. 246.

[5] George M. Hall, *Image Processing: A Management Perspective* (New York: McGraw-Hill, 1991), p. 99.

[6] Ibid., p. 100.

[7] Ibid., p. 101.

[8] Ibid., p. 103.

5

Chapter 5

Imaging Trends

Introduction

Ask experts about the state of imaging today, and they are likely to talk about changing technology, fragmented systems and a small but growing base of experienced users. Ask them where imaging is headed, and they are likely to mention enterprise-wide systems, integration with other applications and simplified software on the desktop for every user.

There's no question that the future of imaging is strong, optimistic and full of potential. When that vision will become a reality is much less certain. Will local government be a part of the future of imaging? Probably, but not all local governments will move into the world of imaging at the same time.

Local governments will have to weigh the cost benefits of the more innovative forms of imaging with the needs of a changing society. Just how practical is voice-annotated imaging on a WAN when potholes in county roads are taking too long to repair? How important is it to tie document images of land records into a GIS database when funding shortfalls are forcing the city to close library branches? Should every government employee with a computer have access to images when there aren't enough textbooks for every student in school? The private sector doesn't have to wrestle with these issues, but local governments do.

Yet, it's likely that some future imaging applications will help cities and counties meet their mission of providing better service at lower costs. The more paperwork managed electronically, the better government's access to information for analysis, decision-making and distribution.

Desktop Imaging

Imaging in local government today is dominated by a small number of mission-critical applications performing very specific tasks. These imaging systems tend to be expensive, require specialized programming and use document management, workflow and mass-storage tools and services.

Some vendors, however, have begun to introduce imaging products that are both inexpensive and easy to use. The new software is strictly PC-based and is often a component of other applications. Pop open a word processing program and there, next to a paragraph referring to a letter from the chamber of commerce to the mayor, is a thumbnail-sized image of the letter. Double-click on that image and it expands to full size for reviewing and editing.

These imaging applications are intended for the occasional user of imaging. They are not meant to replace systems that handle high-volume applications, such as those for land records, police reports, parking tickets and the like. A government is more apt to use desktop imaging for town hall correspondence, a public defender's court cases or documentation for a management report, to name just a few possibilities.

Right now, these new products lack features that more established imaging products include, such as indexing, workflow and mass storage service. But competition could drive down the costs of high-end imaging, while forcing the new line of low-end imaging products to add more features.

What's also missing at this point is a compelling reason to purchase imaging for occasional use. As a stand-alone tool, desktop imaging has limited capabilities because current desktop products lack the software hooks that would enable them to pull images off the powerful high-end servers that manage most of today's imaging applications in local government.

The newer products also lack standards for interoperability with established imaging systems. Without standards, integration between a low-end imaging product and the document management services of a high-end imaging system will remain difficult, if not impossible. That, in turn, makes low-end imaging harder to cost-justify, despite its attractive sticker price. If desktop applications could allow users to share images stored in other imaging systems, they would be far more powerful. Fortunately, at least one group of imaging and document management professionals—known as the Shamrock Group—has been working since 1993 to harmonize the different kinds of imaging products available today and in the near future.

Future Imaging Applications in Local Government

Despite the current lack of tools for integrating low-end imaging software with established, high-end imaging applications, integration will play a key role in the future of imaging. Leading imaging vendors are developing tools that share images across applications.

Integration of imaging and geographic information systems (GIS) is one of the most promising data "marriages" for local governments. Until recently, use of imaging with GIS was largely limited to satellite imagery and aerial photography, overlaid on GIS maps to help identify specific structures, types of land formations and variations in vegetation.

Today, however, applications can link a database of document images with geographic addresses on an electronic map. Take land records and GIS, for example. A single query by a user summons an electronic map that displays the vacant lots in a given section of a city. By clicking on one of the lots, the user can retrieve and display images of the actual land records tied to that lot. Such linkage of document images with geographic locations can be applied to virtually any sector of government operations.

Imaging can also support another growing field of government technology, namely electronic service delivery. Touchscreen kiosks in malls and stores across the country, for example, now provide local government information and services to citizens. Some allow citizens to order copies of birth certificates and hunting licenses or renew a car registration.

Right now, kiosk customers must wait several days for the requested certificate or registration to arrive in the mail. Improved imaging technology may change that entirely. Conceivably, images of these documents could be transmitted from central computers to the kiosk in the mall, where a laser printer could print copies for the customer to take home.

Changing Technology

Besides branching out into other computer applications, such as GIS and multimedia, imaging technology is also extending into other fields. One such field is electronic fingerprinting, which incorporates the scanning and matching of human fingerprints. What used to take law enforcement officers days or weeks to do is now done in seconds, thanks to a computer's ability to compare the image of a newly scanned fingerprint with a database of known fingerprints and to find a possible match.

Besides helping law enforcement agencies identify suspects, electronic fingerprinting enables social service agencies to detect welfare fraud. In several counties nationwide, electronic fingerprinting systems are used to capture an image of each welfare applicant's fingerprint. This database of prints, coupled with the rapid matching ability of the system, acts as a powerful deterrent to individuals hoping to double-dip for benefits.

As OCR technology improves, some software firms are expanding the capabilities of text retrieval systems to include intelligent searches of text and data for specific information. Such searches take advantage of OCR's unique ability to convert scanned documents into searchable text.

Armed with a tool called fuzzy searching, these new applications use keyword criteria to create summarized reports for management executives and analysts. Fuzzy searching can comb through large quantities of text, find articles and reports related to the search terms being used, and create a summary of those documents. That summary and the scanned documents are linked, so that if an individual wishes to see the actual information on which the summary is based, he or she simply clicks on a "hot" key to bring up the document.

Fuzzy searching falls just on the perimeter of artificial intelligence. So do workflow modeling programs. These emerging tools use variations on expert system technologies to help an organization evaluate its current workflow situation and the impact certain changes will have on its work process. The tools can even tie an organization's current costs to the benefits of workflow redesign, allowing a department to test the waters, so to speak, without getting wet.

Document Management

As uses of imaging grow and evolve, it becomes increasingly clear that the technology is no longer limited to single purpose, "line-of-business" applications, but has become one component in a broader document management environment. Document management includes imaging, text retrieval and workflow, as well as spreadsheets, word processing and electronic mail. The data these various applications handle is changing, too. Besides database information and text, applications work with graphics of scanned images and audio and video clips.

The computer of the future will provide users with information they can watch and hear as well as read. Imaging will be a part—but an integral part—of this multimedia approach to information processing. What sort of architecture and standards will support that approach is still unclear.

As a local government prepares for the document management environment of the future, it must not overlook personnel considerations. While computer applications have become easier to operate, thanks to graphical user interfaces and better-designed software, the cultural impact of operating in a near total electronic environment can be overwhelming. Proper training and an understanding of human concerns about jobs and tasks should receive as much attention as system development. Without a well-trained, motivated staff, even the best-designed system will fail.

Preparing for the Future

While local governments don't have to rush to revamp computer systems so that they are capable of running multimedia, they would do well to think of the long-range implications, and applications, of imaging. A few years ago, workflow didn't matter in government imaging applications because the tool simply didn't exist. Today, many versions of workflow are available, and they are becoming easier and easier to use. It would not be surprising if the obstacles that currently prevent imaging from appearing on every worker's desktop disappear in a few short years.

Imaging has developed rapidly in its short existence. Any local government venturing into imaging today should keep a close eye on the technology to see where it is headed and be prepared for the coming changes. How does one plan for the future?

- **Read the literature.** A number of publications cover imaging, information technology in local government and both.

- **Attend conferences and workshops.** Trade shows are a valuable way to see firsthand the latest in technology; workshops offer many opportunities to pick up expert advice.
- **Network.** The number of imaging users in local government is still small. Find out who some of these innovative few are and stay in touch with how they are doing. If any are within traveling distance, find out if staff would be willing to give a demonstration of the jurisdiction's imaging system. Many early innovators are proud of their achievements and willing to explain them.

The technology of imaging will eventually come to virtually every local government, large and small. Unlike computer systems of the past, which merely automated existing processes, imaging can truly change the way work is performed and services are delivered. It may transform the way cities and counties govern. With proper planning, sound management and effective leadership, local governments can take advantage of imaging and the opportunities it can unleash.

Appendix A

Imaging Strategic Plan—Hennepin County, Minnesota

October 2, 1992

Hennepin County Information Services

Table of Contents:

Overview

The purpose of this document is to assist decision makers who are planning and implementing imaging systems at Hennepin County over the next three to five years. This will be done by:

1. Providing background information on problems to be solved, imaging system solutions, and the strategies for getting there
2. Identifying Information Services goals and commitments to support imaging system implementations
3. Describing both existing and planned county application systems using this technology

The County Board and Administration have recognized that imaging technology can help provide solutions to increasingly burdensome document management problems. Late in 1991, Information Services was directed to become familiar with this technology in order to assist county departments with the acquisition of solutions. Early in 1992, Information Services assigned a full-time systems analyst to begin learning. The county became a partner in the Minnesota Imaging Project, a multicompany research effort. In July 1992, a team was established for the purpose of developing and maintaining a strategic plan. The team includes:

Anne Collopy, HCIS Records Management
Larry Ingram, HCIS Technical Services
Jackie Weiler, HCIS Development
Gary Yochum, HCIS Development

This plan was drafted by Information Services with valuable assistance from other departments. Community Corrections, Community Services, County Recorder, Medical Center, Probate Court, and the Sheriff's Department are areas that have already invested a lot of time looking at this technology. Their knowledge of imaging and commitment to effective use of this technology must be recognized.

This plan will be updated periodically by HCIS for use on an ongoing basis.

Background

The Problem

Hennepin County, like most large government jurisdictions, is heavily involved in the business of creating *records*—records on its citizens, history, lands, and institutions. Until the end of the 1960s, county agencies were overwhelmingly paper-intensive; even the advent of computer technology and advanced microfilming techniques did little to reduce the burgeoning stockpiles of "hard copy," and indeed, in many cases, added to them. For instance, computer printouts and documents are often retained even after having been filmed.

During the 1970s and continuing to the present day, Hennepin County departments have undertaken many efforts to reduce paper output and storage. Among them are:

- Records retention schedules
- Improved filing systems and equipment
- More sophisticated microfilming policies and practices (including use of archival microfilm and destruction/recycling of paper documents)
- Well-managed off-site storage programs
- On-line report viewing

However, much remains to be done; great progress can be made through the judicious application of still newer technologies, including imaging.

The very nature of most governmental functions makes imaging an attractive if not a compelling approach to records and document management problems. For one thing, the creation of government records is usually mandated by state or federal law. The interaction of citizens with government agencies—always resulting in added records—is seldom voluntary (for example, one cannot obtain a driver's license, resolve a court matter, or apply for various social welfare benefits without going through government agencies). Moreover, there is often a need for sharing of records between agencies, such as in law enforcement. Records multiply and records are lost. Records proliferate and records need to be protected. Records are vital to the county: what affects them affects citizens in direct, personal, and sometimes dramatic ways.

Imaging as a Solution

Imaging is the ability to electronically capture, store, retrieve, display, process, and distribute business objects. At this time, the primary area of opportunity is documents; however, many of the same strategies will apply with other objects such as voice and video.

Imaging systems can generally fit into one or more of these categories:

- Basic storage and retrieval
- Workflow—objects are routed to one or more work queues in a parallel or sequential fashion
- Folder—objects are stored in "folders" with tabs and subtabs

Imaging systems require hardware and software that function on at least one, and usually two or more of the following platforms:

- Stand-alone workstation
- Local area network with workstations and server(s)
- Mid-range systems
- Mainframe systems

Typical components in an imaging system are described here:

Workstation. Personal computer operated to capture or request a document. At a minimum, the workstation would include a CPU, monitor, and imaging software. Scanner, printer, fax capabilities, magnetic storage, and/or optical storage may be attached depending on requirements.

Monitor. High resolution device that displays the image. It can be separate from the data system monitor or it can be the same one. The screen size and resolution of the display vary depending on the requirements.

Scanner. Input device that reads images and digitizes them so they can be electronically stored. Many different types are available including: hand held, flatbed, auto feed, curved bed, one side scanned, both sides scanned, etc. The size, type, and numbers of the documents to be scanned generally determine the type and cost of the scanner.

Printer. Device used to produce a paper copy of the image.

FAX. Low cost input and output device used with many imaging systems.

Optical Storage. Primary storage device for images. For the same size disk, optical has about 100 times the capacity of magnetic storage, but retrieval is slower.

Jukebox. Device that contains multiple optical disks that are mechanically loaded and unloaded onto one or more read-and-write devices called drives.

Image Server. Processor that manages the storage and retrieval of images. This may also function as a workstation on a small system. However, it is usually a dedicated

PC, mid-range system, or mainframe. The image server is typically accessed over a local area network by "client" workstations.

Indexes. Data files created to allow the effective retrieval of images. Careful consideration must be taken when designing the indexing scheme. The indexes are usually stored in a data base on magnetic storage separate from the imaged objects.

This technology presents obvious opportunities for county departments to improve service and reduce costs. Many departments have been actively pursuing information regarding imaging systems being implemented in similar business situations outside the county. Some have also worked with vendors in attempting to find solutions to their own business problems. The implementation has already started within the county at the Medical Center and the Sheriff's Department. It is expected that many more departments will follow over the next five years. The applications will be described later in this document.

Costs. Hardware, software, installation, training, and maintenance all must be considered in the cost of an imaging system. Initial hardware and software costs range from under $10,000 for a single workstation system to several million dollars for a system with 500 workstations with access to one or more jukebox storage devices. The costs for both workstations and storage have been dropping, and that trend is expected to continue. This is expected to increase the number of applications that will be cost justified. The following list provides approximate costs for imaging system components.

$1,500	Image retrieval capabilities added to an existing workstation
$6,200	Workstation with monitor and imaging software
$150–100,000	Scanner
$500–10,000	Single optical drive
$10,000–500,000	Jukebox
$1,000–20,000	FAX server hardware and software

Image server hardware and software:

$125,000	Mainframe
$50,000–100,000	Mid-range
$10,000	LAN-based

Benefits. Imaging systems now on the market offer advantages including but by no means limited to:

- Extremely high density storage of massive quantities of data
- Ability to unify divergent types of records into a single file, and to access part or all of that file quickly
- Protective backup for paper documents
- Wide dissemination and simultaneous use of records
- Admissibility of "imaged" records as evidence in all courts and legal proceedings (Minnesota Statutes §§15.17, Subd.1 and 138.17, Subd.1)

While imaging technology is relatively new and costly and must never be considered a panacea, there are enough records concerns in Hennepin County that it might be employed with impressive results in selected offices. This strategic plan has been prepared with such potential payoffs in mind.

Tangible benefits often include:

- Faster retrieval of documents
- Elimination of lost files
- Simultaneous access to a single document

- Reduced physical storage space
- Simplified processing due to better workflow or control
- Staff reduction

Other benefits more difficult to measure include:

- Improved overall work environment and morale
- Increased document security
- Improved customer service

Challenges. Organizational and technical challenges will present barriers to the successful use of imaging. Some of these challenges are unique to this technology while others are common when introducing change to a diverse and complex organization such as Hennepin County.

It is often necessary to change the way an organization does business to take best advantage of imaging. Current workflows need to be reviewed and often revised. This can be very difficult for some organizations to implement.

Even at this early stage, the costs and complexities of imaging as a solution are apparent. It will be expensive, some mistakes will be made, and some conversions will be needed. It will go too fast for some people and not fast enough for others. While the technology has come far, it is not yet sufficiently advanced to meet all county requirements.

Vendors are offering many products with promises of solutions to many problems. The products and technology are changing quickly in function and cost. Some vendors offer solutions that meet isolated business requirements but lack compatibility for integration with other business areas or systems. These factors present risks and point to the need for ongoing efforts to monitor changes and their impact on county systems. There are few standards across vendors or even across products from a single vendor. One problem is that different algorithms are used for data compression and storage. That can make it difficult to pass images between systems.

Images are stored as binary large objects (BLOBs). The compressed image of a single 8 1/2" X 11" black and white document is stored as a 50,000 byte BLOB. The same document if keyed could be stored in about 2,000 bytes. Delivering such large amounts of data to a workstation with an acceptable response time presents new opportunities in managing system performance.

Requirements for each application must be gathered in order to determine the appropriate hardware and software platform to best meet the application needs. Hardware or software platform bias can cloud the selection process. There are a variety of vendors that support county systems. It may be that an organization would prefer to work with (or not work with) a particular vendor depending on prior experience. This vendor bias can lead to selecting an imaging solution that may not be the most appropriate. While these challenges can be serious, awareness and planning can minimize their impact on system implementations.

Getting There

The costs are high, the benefits are higher, and the challenges are significant. The vision of multiple image-enabled business systems sharing images across departments as needed is the long-range goal. Getting there requires both strategic planning and tactical planning as projects acquire and implement individual systems.

Strategic planning. Effective implementation of this technology across county departments depends on informed decisions being made with an understanding of how those decisions fit into future plans. A common vision of where we want to be and strategies for getting there are important. The remainder of this document describes the strategic goals and commitments that HCIS is making to support the effective use of imaging technology.

Projects. Projects that deliver image-enabled business systems will provide valuable experience as the county takes advantage of this emerging technology. To ensure our opportunity to learn from experience, projects should follow a process that includes:

1. Defining business problems, processes, and the expected benefits of the imaging system
2. Defining detailed system functional, technical, and performance requirements
3. Selecting hardware and software that meet system requirements and fit within county standards and guidelines

The Information Services Development Division Application Groups are preparing to provide project support in analysis, design, and implementation of imaging systems. Initial projects may have limited scope but will provide experience in particular business areas. That experience is expected to provoke ideas for expanded uses of the technology.

Strategic Goals

HCIS has identified four major goals to support county acquisition of imaging technologies. The goals are defined in greater detail in the remainder of this document. Over the next five years HCIS, working with other county departments, will attempt to

1. Determine general and application-specific county imaging requirements
2. Document hardware and software platform assumptions and alternatives that may impact imaging
3. Educate county management and staff
4. Maintain a profile of installed imaging applications and plans for those anticipated within the next five years

Meeting these goals will promote sharing of information and experience between departments. It will also provide a way to share long-range plans. This sharing will help to ensure that future applications can share both the images and the index data used to identify them.

County Requirements

This section expands on the first strategic goal: *HCIS, working with other county departments, will attempt to determine general and application-specific county imaging requirements.*

Using imaging, or any other technology, to solve business problems requires clearly defined requirements. Each business system has unique requirements. Imaging systems usually require some degree of integration with other automated systems. Following the steps described in this section will help project teams determine requirements for an imaging system and then design and acquire a system that both delivers the functions required by the business and fits within the county's planned technical environment.

Determining requirements requires project teams to answer questions relating to such items as number of documents, frequency of access, retention, indexing, security, and workflow. Specifications can be defined for the new system. The specifications should include detailed descriptions of inputs, processing, outputs, hardware, and other technical requirements for the system. These specifications become the basis for a Request for Proposal, purchases, and/or a system development effort.

When developing specifications, consideration must be given to both the current and future technical environment in which the system will operate. The next section in this document describes those environments as seen today. Technical environment standards will serve as a baseline for new systems. As with any standards, exceptions will at times be necessary. It is important, however, to document exceptions so they can be considered in future planning.

To further define the steps summarized above, Information Services has developed the *Imaging Specifications Guideline*. This document contains:

1. A detailed "Requirements Interview Checklist" to assist in gathering requirements for imaging systems
2. Specifications for functions that will be common across many county applications

The document can be obtained from Jackie Weiler at 612/348-4995.

Technical Assumptions and Alternatives

This section expands on the second strategic goal: *HCIS, working with other county departments, will attempt to document hardware and software platform assumptions and alternatives that may impact imaging.*

Understanding both the current and future technical environment in which imaging systems will operate is critical to those doing projects and those making decisions at a high level. This is necessary, however, due to the high costs in acquiring this technology and the critical dependence on these systems by county business areas. This section describes those environments as seen today, assumptions about the future, and alternatives for consideration where appropriate.

Because of constant change in many high technology areas that impact imaging systems, it is difficult to describe the present technical environment and even more difficult to describe the future. The following should be considered:

Workstation hardware and operating systems. The county has been purchasing IBM PS/2 models almost exclusively and is expected to continue for the foreseeable future. Apple Macintosh workstations are used in some situations.

Server platforms. The county has been purchasing the larger IBM PS/2 systems to provide file and print server functions. Those servers are based upon the 0S/2 operating systems. Future servers, particularly image servers, may require other platforms.

Communications networks. The county has invested significant dollars and effort to construct a network to support current needs and to provide for future growth in numbers of installed workstations and data transmission needs (image and otherwise). The current network is based on IBM token-ring network technology over either IBM type-1 or fiber optic cable. It currently operates at either 4 or 16 megabytes but can be easily upgraded to 100 megabytes transmission speeds.

Graphical user interfaces. Graphical user interfaces, in their many flavors, will be the user interface of choice. IBM SAA CUA, 0S/2 presentation manager, and the like will be the normal application user interface by the mid 1990s.

Electronic mail. Within the short term, the county will continue to provide e-mail services via the central mainframe EMC2 system. As the county extends the token-ring network and, more important, replaces single function terminals with PC workstations, a LAN-based electronic mail facility may provide improved functionality and ease-of-use.

Image mail. The ability to link written word messages, spoken word messages, full motion video, document images, and to "mail" these items is an emerging technology. It could be expected that the technology to provide a practical implementation of image mail will be a future benefit of most image systems.

Client/server strategies. The county is starting to develop applications where more than one computer is involved in the completion of a requested process. Imaging systems generally utilize this approach. The client application performs functions such as image compression/decompression and presentation of the user interface—screens, commands, messages, etc. The server application typically manages indexes and images.

Education

This section expands on the third strategic goal: *HCIS, working with other county departments, will attempt to educate county management and staff.*

This goal will be achieved through:

Training. Training will be provided for county staff and management. HCIS business analysts and imaging specialists will be trained to assist departments in acquiring imaging technology.

Sharing of research and experience across departments. This strategic plan will be updated periodically and provided to interested departments. Technology briefings are also available for interested departments. Training presentations and documentation will be provided throughout imaging implementation projects.

Monitoring vendor offerings. Some vendor product profiles are available and will be updated as projects gather additional information.

Monitoring imaging technology and applications outside the county. The county participates in the Association for Information and Image Management (AIIM) and the Minnesota Imaging Project (MIP). These organizations provide valuable contact with vendors and experienced users of imaging technology.

Managing expectations. This document and related project documents will attempt to promote reasonable expectations as imaging technology emerges and evolves.

County Applications

This section expands on the fourth strategic goal: *HCIS, working with other county departments, will attempt to maintain a profile of installed imaging applications and plans for those anticipated within the next five years.*

Existing

Currently, two imaging systems are in use at Hennepin County.

- A contracted digital imaging service is used by the Medical Center's Medical Imaging Division. Approximately 300,000 microfilmed X-rays were digitized and archived between March 1991 and March 1992, with $132,000 budgeted. Eight prior years of filmed X-rays have now been converted. Over 500,000 images are estimated to be done in 1992-1993. Images are currently available for viewing and printing at a small number of workstations. The system has been so successful and well received that there is no doubt it will continue to grow.
- The Sheriff's Department recently installed a fingerprint system. The system uses two stand-alone workstations, each with a scanner and a printer attached. The system is expected to speed up the booking process.

Potential

Potential applications of imaging technology might be limitless except for budgetary constraints. Undoubtedly, within 10 years applications which today are prohibitively expensive will be fully operational; and systems now undreamed-of will be used as models for other counties.

Following are several of the more promising possibilities:

- The County Recorder's office is a possible candidate for imaging. Its Tract Index is already an on-line IMS application, and the existing document numbering system could be used for indexing. This business area has been image enabled by other counties.
- Probate Court has investigated imaging technology for two years, including a workflow study done with the Office of Planning and Development. They are nearly ready to make a serious commitment. Their proposed initial system would accommodate all new probate and trust cases on a day-forward basis; would require two scanners, seven or eight display terminals, and four or five printers; and would probably interface with a state system.
- Community Corrections is eager to begin imaging. They want to create client-based folders of their Adult Field Services Investigation and Supervision Divisions files. Probation Officers could search one source for all information regarding an offender; this would include remote access by suburban courts. (The current system is in some disarray, with long delays in obtaining necessary information for Pre-Sentence Investigations and other critical functions assigned to these divisions.) The imaging system would be integrated with the new data system currently being developed.
- Environmental Management Divisions, such as Solid Waste and Hazardous Waste, would be greatly helped by the ability to store their countless federally mandated forms and records on optical disks. The current system is drowning in paper. The potential for improved workflow management exists here, too.
- Voluminous Purchasing specifications and contracts, particularly for standard services and commodities, might be better "imaged" than stored in paper.
- Workers' Compensation claim files, retained "forever," have many features which would lend themselves to an imaging system. Indeed, the State of Minnesota already has made the conversion.
- The vision of Hennepin County Medical Center for patient care is to provide a more totally integrated health care delivery system. This includes developing outpatient clinics at different locations throughout the county. The vision for Health Information Management is to evolve from a paper-based patient medical record to a completely interactive multimedia-based system. Imaging is one piece of the future patient medical record.

  The current paper medical record is in great demand as a treatment tool, for monitoring quality of care, and as a business document. It is not capable of meeting the timely access needs of all authorized users. As usage grows, the need increases for more controlled access to retain patient confidentiality. An imaging system is an ideal first phase in realizing this vision. It will improve record availability and control authorized access to the patient's medical record while increasing user efficiency. It is estimated that Hennepin County Medical Center will require approximately 1,000 points of entry into an imaging system to be functional. This gives an idea of the magnitude of the application.
- Economic Assistance case files pose an opportunity for workflow improvement with an imaging system. Even though cases may be destroyed legally four years after closing, many clients come in repeatedly, bringing dependent family members and participating in several programs over decades. As folders thicken with countless federal and state-mandated forms, new "volumes" are created—for example, a case could have two or three volumes (or more), with material in Volume I being 15-20 years old. In 1991, Economic

Assistance received permission from the state to destroy nonessential records older than four years in order to cope with the ongoing filing space crisis. Even with that unusual concession, they remain buried under a blizzard of paper. Access to client i.d.s with photos by food stamp issuance offices would reduce a number of problems currently being experienced.

- Community Services applications are similar to those of the HC Medical Center; that is, there is interest in migrating from a paper-based client record (medical and social services) to an interactive image capture system. Of particular merit would be the movement of data from inactive records storage to authorized workers in real-time operations. Currently, over 150,000 inactive records are stored in various records centers supported by the department.

  Community Services also suffers from the concerns posed by the EA Department. Client records bulge with countless federal and state mandated forms. Key examples include case records at the Detoxification Center and case records in the Developmental Disabilities Division.

  The department currently has extensive use of microfilm for case records which require permanent storage (e.g., adoption records), and for the storage of client billing transaction forms, financial determination forms, and other source documents which are keyed by staff for accounts receivable and accounts payable processing of client accounts.

  Non-clinical areas, specifically accounting and areas within the Community Resources Division, have a myriad of potential workflow productivity improvement applications where there is a need for repetitive look-up of static information by multiple workers. Examples include vendor Purchase of Service contracts, policy and procedural manuals, resource and referral listings, etc.
- Photo identification systems have been looked at by Property Management, Community Corrections, and the Medical Center.
- The Sheriff's Department is currently in the process of obtaining a mug shot system.

Law enforcement and County Attorney functions certainly must be included on any list of imaging candidates, although each of these areas presents specific needs—and specific difficulties—that might suggest postponement of anything like a full-fledged system. The county should continually appraise industry-specific imaging systems for law enforcement documents and closely follow what other jurisdictions and agencies are doing with booking records and Adult Prosecution (criminal) case files, etc.

The following schedule is intended for strategic planning purposes only. It does not imply the existence of either a tactical plan or funding to accomplish the implementation.

Imaging Workstation Projections by 1997

	Year	Workstations
HCMC Imaging Division X-rays	1991	1
Sheriff—Fingerprints	1992	2
County Recorder's Office		
Abstract	1993	9
Torrens		
Probate Court		
Phase I	1993	7
Phase II	1994	
Phase III	1995	
Community Corrections		
Phase I	1993	32
Phase II	1994	200
Environmental Management		
Purchasing		

	Year	Workstations
Personnel—Worker's Compensation		
HCMC—Medical Records	1993	50
		950
Economic Assistance		1200
Community Services		
Worker Control/Billing	1992	1
Activity Forms		
Other Applications		600
Total by 1997		3000+

It is very difficult to predict accurately the speed at which this technology will be implemented, or the costs and benefits to be realized by the county. Industry experts project that costs for workstations and storage will continue to fall. As costs fall, more systems will be justified. Basic planning assumptions include the following:

- It is likely that within a few years, the cost of adding image retrieval capabilities to an existing workstation will drop to well under $1,000, down from a current cost of about $ 1,500. This does not include the costs of image storage or access.
- Images will be stored on different media such as magnetic disk, optical disk, and tape.
- Image storage will eventually be attached to multiple computer platforms such as individual workstations, local area networks, mid-range, and mainframe.
- 1992 costs per image-enabled workstation average about $22,000. This includes the workstation itself, attached scanner and/or printer as required, and a portion of an image server with attached storage. Averages can be very misleading however, because the range for an individual workstation goes from under $4,000 to over $30,000, depending on hardware already available, hardware and software added, and image storage.
- Costs for imaging system components will continue to fall. Costs per workstation are generally lower for large systems as fixed costs are spread across more workstations.
- Current projections indicate that over 3,000 workstations will be capable of displaying images within five years. Many of these workstations will already have been in use for word processing and other applications.

The previous assumptions and schedule are intended to provide an early vision of what might occur.

Appendix B

Standards and Procedures for Electronic Records of Local Governments

Texas Administrative Code
Title 13, Chapter 7

§7.71. Definitions.
The following words and terms, when used in this chapter, shall have the following meanings, unless the context clearly indicates otherwise. For local governments, terms not defined in these rules shall have the meanings defined in the Local Government Code, Title 6, Subtitle C, Chapter 201. For state agencies, terms not defined in these rules shall have the meanings defined in the Government Code,§§441.031-441.039 and §§441.051-441.062.

AIIM–The Association for Information and Image Management.

ANSI–The American National Standards Institute.

Archival record–A record of a state agency scheduled to be reviewed by or that has been approved by an archives for permanent preservation.

Database–(A) collection of digitally stored data records; (B) collection of data elements within records within files that have relationships with other records within other files.

Database Management System (DBMS)–Set of programs designed to organize, store, and retrieve machine-readable information from a computer-maintained database or data bank.

Data file–Related numeric, textual, sound, or graphic information that is organized in a strictly prescribed form and format.

Electronic media–All media capable of being read by a computer including computer hard disks, magnetic tapes, optical disks, or similar machine-readable media.

Electronic record–Any information that is recorded in a form for computer processing and that satisfies the definition of a state record in the Government Code, §441.031(5), or the definition of local government record data in the Local Government Code, §205.001.

Electronic records system–Any information system that produces, manipulates, and stores state or local government records by using a computer.

IEC–International Electrotechnical Commission.

ISO–International Organization for Standardization.

Long-term record–A record for which the retention period on a records retention schedule is 100 years or more but less than permanent.

Medium-term record–A record for which the retention period on a records retention schedule is 10 years or more but less than 100 years.

Records administrator–The person appointed by the head of each state agency to act as the agency's representative in all issues of records management policy, responsibility, and statutory compliance.

Records custodian–The appointed or elected public officer who by the state constitution, state law, ordinance, or administrative policy is in charge of an office that creates or receives local government records.

Records management officer–Each elected county officer or the person designated by the governing body of each local government pursuant to the Local Government Code, §203.025.

Short-term record–A record for which the retention period on a records retention schedule is less than 10 years.

Permanent record–A record for which the retention period on a records retention schedule is permanent.

Text documents–Narrative or tabular documents, such as letters, memorandums, and reports, in loosely prescribed form and format.

§7.72. General.

(a) These rules establish the minimum requirements for the maintenance, use, retention, and storage of all medium-term, long-term, and permanent electronic records of state agencies and local governments, and archival electronic records of state agencies. These rules do not apply to short-term electronic records, but the short-term electronic records of local governments are subject to the applicable provisions of the Local Government Code, Chapter 205.

(b) Unless otherwise noted, these requirements apply to all electronic records storage systems, whether on microcomputers, minicomputers, or main-frame computers, regardless of storage media.

(c) An electronic storage authorization request certifying that these rules will be followed must be submitted to and approved by the director and librarian for all existing electronic storage of medium-term, long-term, and permanent state or local government records and state archival records. The authorization request must be submitted in a form and manner to be determined by the director and librarian and must be signed by the agency head or designated records administrator (for state agencies), or the records management officer (for local governments).

(d) The agency head or designated records administrator (for state agencies), and the governing body or records management officer in cooperation with records custodians (for local governments) must:

(1) administer a program for the management of records created, received, maintained, used, or stored on electronic media;
(2) integrate the management of electronic records with other records and information resources management programs of the agency;
(3) incorporate electronic records management objectives, responsibilities, and authorities in pertinent agency directives;
(4) establish procedures for addressing records management requirements, including recordkeeping requirements and disposition;
(5) ensure that training is provided for users of electronic records systems in the operation, care, and handling of the equipment, software, and media used in the system;
(6) ensure the development and maintenance of up-to-date documentation about all electronic records systems that is adequate to specify all technical characteristics necessary for reading or processing the records and the timely, authorized disposition of records; and
(7) specify the location and media on which electronic records are maintained to meet retention requirements and maintain inventories of electronic records systems to facilitate disposition.

(e) With the exception of subsections (c) and (f) of this section, which are effective immediately, state agencies and local governments must be in compliance with the Standards and Procedures for Electronic Records on or before January 2, 1995.

(f) Any electronic recordkeeping system not meeting the provisions of these rules may be utilized for medium-term, long-term, or permanent state or local government records and state archival records provided the source document, if any, or a paper copy is maintained, or the record is microfilmed in accordance with the specifications in *American National Standard for Imaging Media (Film)—Silver-Gelatin Type—Specifications for Stability* (ANSI IT9.1-1989 or latest revision) for state records or in accordance with the provisions of Local Government Code, Chapter 204, and the rules adopted under it for local government records.

§7.73. Creation and Use of Data Files.

(a) Disposition instructions for the data must be incorporated into electronic records systems that produce, use, and store data files.

(b) State agencies and local governments must maintain up-to-date technical documentation for each electronic records system that produces, uses, and stores data files. Minimum documentation required is:

(1) a narrative description of the system;
(2) the physical and technical characteristics of the records, including a record layout that describes each field including its name, size, starting or relative position, and a description of the form of the data (such as alphabetic, zoned decimal, packed decimal, or numeric), or a data dictionary, or the equivalent information associated with a database management system including a description of the relationship between data elements in databases; and
(3) any other technical information needed to read or process the records.

§7.74. Creation and Use of Text Documents.

(a) Electronic records systems that maintain the official file copy of text documents or data used to generate the official file copy of text documents on electronic media must meet the following minimum requirements:

(1) provide a method for all authorized users of the system to retrieve desired documents, such as an indexing or text search system;
(2) provide security to ensure integrity of the documents;
(3) provide a standard interchange format when determined to be necessary by the agency or local government to permit the exchange of documents on electronic media among the components of the agency or local government using different software/operating systems; and
(4) provide for the disposition of the documents including, when necessary, the requirements for transferring archival records to the State Archives as detailed in §7.77 of this title (relating to Retention of Electronic Records).

(b) A document created on an electronic records system must be identified sufficiently to enable authorized personnel to retrieve, protect, and carry out the disposition of documents in the system. Agencies must ensure that records maintained in such systems can be correlated with related records on paper, microform, or other media.

§7.75. Security of Electronic Records.

(a) State agencies and local governments must implement and maintain an electronic records security program for office and storage areas that:

(1) ensures that only authorized personnel have access to electronic records;
(2) provides for backup and recovery of records to protect against information loss;
(3) ensures that personnel are trained to safeguard confidential electronic records;
(4) minimizes the risk of unauthorized alteration or erasure of electronic records; and
(5) documents that similar kinds of records generated and stored electronically are created by the same processes each time and have a standardized retrieval approach.

(b) A duplicate copy of essential records and any software or documentation required to retrieve and read the records must be maintained in a storage area located in a separate building from the building where the records that have been copied are maintained.

(c) For all permanent records stored on rewriteable electronic media, the system must ensure that read/write privileges are controlled and that an audit trail of rewrites is maintained.

§7.76. Maintenance of Electronic Records Storage Media.
(a) State agencies and local governments must ensure that the accuracy, completeness, and accessibility of information are not lost prior to its authorized destruction date because of changing technology or media deterioration, by converting electronic storage media and taking other action as required to provide compatibility with current hardware and software. The migration strategy for upgrading equipment as technology evolves must be documented and include:

(1) periodically recopying to the same electronic media as required, and/or transferring of data from an obsolete technology to a supportable technology; and
(2) providing backward system compatibility to the data in the old system, and/or converting data to media that the system upgrade and/or replacement can support.

(b) Paragraphs (1)–(3) of this section outline the maintenance of backup electronic media stored offsite.

(1) Magnetic computer tapes must be tested and verified no more than 6 months prior to using them to store electronic records. Pretesting of tapes is not required if an automated system is used that monitors read/write errors and there is a procedure in place for correcting errors.
(2) The storage areas for electronic media must be maintained within the following temperatures and relative humidities:
    (A) for magnetic media—65 degrees Fahrenheit to 75 degrees Fahrenheit, and 30% to 50% relative humidity;
    (B) for optical disks—storage environmental conditions as specified in *Information technology—130 mm optical disk cartridge, write once, for information interchange* (ISO/IEC 917-1, 1990 or latest revision).
(3) A random sample of all magnetic computer tapes must be read annually to identify any loss of data and to discover and correct the causes of data loss. At least a 10% sample or a sample size of 50 magnetic tapes, whichever is less, must be read. Tapes with unrecoverable errors must be replaced and, when possible, lost data must be restored. All other tapes which might have been affected by the same cause (i.e. poor quality tape, high usage, poor environment, improper handling) must be read and corrected.

(c) State agencies and local governments must recopy data maintained on electronic media according to the following schedule.

(1) Data maintained on magnetic tape must be recopied onto new or used tape a minimum of once every three years.
(2) An alternative option for recopying magnetic tape is for the data to be recopied onto new tape a minimum of once every ten years, provided the tape is rewound under controlled tension every three and one-half years. The requirement for rewinding does not apply to 3480-type tape cartridges.
(3) Data maintained on optical disks must be recopied a minimum of once every 10 years.

(d) Floppy disks (diskettes) or any type of flexible disk system may not be used for the exclusive storage of medium-term, long-term, or permanent records and state archival records.

(e) External labels, or an eye-readable index relating to unique identifiers, for electronic media used to process or store electronic records must include the following information:

(1) name or other identifier of the organizational unit responsible for the records;
(2) descriptive title of the contents;
(3) dates of creation and authorized disposition date;
(4) security classification;
(5) identification of the software (to include specific application if appropriate) and hardware used; and
(6) system title, including the version number of the application.

(f) Additionally, the following information must be maintained for electronic media used to store permanent electronic records:

(1) file title(s);
(2) dates of coverage;
(3) the recording density;
(4) type of internal labels;
(5) volume serial number, if applicable;
(6) the number of tracks;
(7) character code/software dependency;
(8) information about block size;
(9) sequence number, if the file is part of a multi-media set; and
(10) relative starting position of data, if applicable.

(g) The following standards must be met for electronic records stored as digital images on optical media.

(1) A non-proprietary image file header label must be used, or the system developer must provide a bridge to a non-proprietary image file header label, or the system developer must supply a detailed definition of image file header label structure.
(2) The system hardware and/or software must provide a quality assurance capability that verifies information that is written to the optical media.
(3) Periodic maintenance of optical data storage systems is required, including an annual recalibration of the optical drives.
(4) Scanner quality must be evaluated based on the standard procedures in *American National Standard for Information and Image Management—Recommended Practice for Quality Control of Image Scanners* (ANSI/AIIM MS44-1988 or latest revision).
(5) A visual quality control evaluation must be performed for each scanned image and related index data.
(6) A scanning density with a minimum of 200 dots per inch is required for recording documents that contain no type font smaller than six point.
(7) A scanning density with a minimum of 300 dots per inch is required for engineering drawings, maps, and other documents with background detail.
(8) The selected scanning density must be validated with tests on actual documents.
(9) The use of the Consultative Committee on International Telegraphy and Telephony (CCITT) Group 3 or Group 4 compression techniques is required for document images without continuous tonal qualities. If use of a proprietary compression technique is unavoidable, the vendor must provide a gateway to either Group 3 or Group 4 compression techniques.
(10) Optical drive systems must not be operated in environments with high levels of airborne particulates.
(11) All aspects of the design and use of the imaging system must be documented, including administrative procedures for digital imaging, retrieval, and storage; technical system specifications; problems encountered; and

measures taken to address them, including hardware and software modifications.

(h) Smoking, drinking, and eating must be prohibited in electronic media storage areas.

§7.77. Retention of Electronic Records.
(a) State agencies and local governments must establish polices and procedures to ensure that electronic records and any software, hardware, and/or documentation, including maintenance documentation, required to retrieve and read the electronic records are retained as long as the approved retention period for the electronic records.

(b) The retention procedures must include provisions for:

(1) scheduling the disposition of all electronic records, according to statutory requirements, as well as related software, documentation, and indexes; and
(2) establishing procedures for regular recopying, reformatting, and other necessary maintenance to ensure the retention and usability of electronic records until the expiration of their retention periods.

(c) State records having archival value and scheduled to be preserved at the State Archives must be transferred to the State Archives as the source document, or printed out on alkaline paper for computer generated information, or on microforms that meet the specifications in *American National Standard for Imaging Media (Film)—Silver-Gelatin Type—Specifications for Stability* (ANSI IT9.1-1989 or latest revision).

§7.78. Destruction of Electronic Records.
(a) Electronic records may be destroyed only in accordance with a records schedule approved by the director and librarian or designee or, in lieu of an approved records schedule, an approved records disposition authorization request.

(b) Each state agency and local government must ensure that:

(1) electronic records scheduled for destruction are disposed of in a manner that ensures protection of any confidential information; and
(2) magnetic storage media previously used for electronic records containing confidential information are not reused if the previously recorded information can be compromised by reuse in any way.

(c) The court ordered expungement of information recorded on an optical Write-Once-Read-Many (WORM) system must be implemented according to the recommendations provided in *Technical Report for Information and Image Management—The Expungement of Information Recorded on Optical Write-Once-Read-Many (WORM) Systems* (AIIM TR28-1991 or latest revision).

§7.79. Public Access to Electronic Records.
An electronic recordkeeping system must not provide an impediment to access to public records.

C Appendix C

RFP Types, Content, Format and Terminology, and Evaluation Tools

RFP Preparation and Announcement

RFP preparation is a time-consuming effort. Staff assigned to help prepare an RFP must expect to spend a large portion of their time on this activity. One of the first tasks is to establish a calendar of events covering all phases of RFP preparation, publication, time allocated to vendors' response preparation, bids evaluation, bidders' presentations, selection of finalists, negotiations and expected date of bid award.

Prebid conferences are sometimes used to offer vendors an opportunity to ask questions related to the project and the bid process itself, to view the site, and to address the unique imaging environment. Such conferences also create an opportunity for project staff to describe the project and key issues, and provide a forum for clarification and discussion of the RFP.

RFP Types

Two different types of RFPs are used for imaging procurement. One emphasizes the desired function; the other emphasizes specifications. Most RFPs are a combination of both. Emphasizing the desired function minimizes the hard technical specifications successful bidders must meet. Some mandatory criteria must be identified to eliminate less responsive bids. Detail is avoided, however, to allow vendors freer rein to make their case and address the solution.

Functional RFPs require more preparation and evaluation time. The evaluating team must be in a position to analyze complex technological issues; therefore, it must include experts. A canned approach to a bid response is never adequate for a functional RFP. A thorough response to the RFP ensures that both parties understand the hard issues and options before a local government awards the bid.

RFPs that emphasize specifications tend to have lengthy lists of features, mandatory or desirable, to be provided by the vendor. Such RFPs reduce the amount of work vendors must put into the response. They also simplify the process of bid evaluation, since a jurisdiction will have fewer alternatives to assess. When the desired product can be very tightly defined, these RFPs are very effective.

RFP Content

An RFP typically covers a wide range of products and services. In addition to imaging hardware and software, an RFP can cover back-file conversion services, database design, applications development and a variety of consulting services.

Local governments must avoid using sample RFPs provided by vendors. This method defeats the very intent of the process. RFP language should also avoid using vendor-specific terminology, descriptions of design, functions, etc. An introduction to an RFP might contain a statement assuring bidders that any use of vendor-specific terminology is accidental.

Although many variations are possible, a typical imaging RFP includes the following sections:

Background information. A section should briefly describe the jurisdiction's imaging process, including any information from previous studies and findings, especially results of the requirements analysis, feasibility studies or other pertinent data. The section should describe the status of imaging planning and list the agencies participating in the process.

Description of jurisdiction. Prospective bidders will need information about the jurisdiction: facts on area size and economy, population and employment, major industries or facilities and municipal services.

Description of existing computing environment. The jurisdiction's existing computing and communications platforms are extremely important to the prospective bidder. The RFP should describe existing resources and supply documentation.

Description of the procurement process. The RFP should include information such as the time and place for the receipt of proposals, how vendor questions will be handled, arrangements (if any) for prebid conferences and names of contacts.

Needs analysis and the implementation plan. Documents from the study phase should be attached to the RFP.

Description of the evaluation process. The RFP should explain evaluation criteria, especially any pertinent facts regarding response to the bid and price evaluation. Evaluation criteria differ widely in practice. They are organized by categories and items, and are weighted. The following are examples of evaluation categories:

- Hardware platforms, operating systems, disk storage and workstations
- Communications
- Scanners, jukeboxes and printers
- Database structure and integration with existing databases and applications
- Function: all required system capabilities and applications
- Future capabilities: hardware, network expansion
- Training
- Support
- Future company direction
- Cost.

The most important criterion on which local governments base their system selection is functionality. Next in importance is the system's ability to integrate with the current computing environment, followed by its ease of use.

Clarification of the role of subcontractors and other suppliers. The complexity of imaging procurement often results in joint vendor proposals. It is important to clarify the roles of bidders, the role of the primary vendor, and the status of other suppliers, as well as the eligibility of vendors to place future bids on products and services related to the same project.

System acceptance. Criteria for system acceptance provide guidelines to bidders as they structure their responses. These criteria should include proposed remedies in the case of vendor non-performance.

Requirements definition. The RFP must also include a description of the solution the jurisdiction seeks, in terms of performance capabilities (functions, reliability, productivity, delivery), configurations, methods and scheduling. The RFP should detail functional and operational requirements and expectations of the system in both short- and long-term contexts. Each characteristic or capability must be measurable. The technical content should be comprehensive yet nonrestrictive. Testing and inspection procedures should be explained.

Documentation. The section on documentation outlines the vendor's responsibility to provide adequate documentation both for the RFP evaluation process and for performance under the contract. For the former, samples of various manuals are useful; for the latter, such items as studies and status reports are valuable.

Contractual requirements. Contractual requirements describe the terms and conditions under which the jurisdiction will purchase a product or service. They include standard items such as scope of work, contract administration, performance guarantees and notice regarding delays.

Bidder profile. The section on bidder profile should request information about a vendor's approach, its imaging experience and the imaging support it provides (e.g., marketing, research and development, customer support); a list of project personnel; references; information on existing clients; a description of third-party support (e.g., for hardware, software or consulting); a description of any planned product developments; and a financial statement.

Pricing. Instructions regulating the pricing of the proposal are critical in imaging procurement. The diversity of packages and functions makes it extremely difficult to compare competing offers. Therefore, it is important to require that all items be priced separately and that the vendor provide prices for database development, database modeling and selected applications. If the imaging hardware and software vendor is not the supplier of database consulting services or the applications software, then the vendor or a consultant will provide estimated costs for this work. When the imaging vendor or a consultant provides these services, the local government must clarify whether these estimates are binding or whether they are subject to reevaluation during the negotiation stage. This method of price comparison is the only equitable process available. It will be generally unfair to vendors who provide more functions to be compared with others on a lowest-common-denominator basis.

Format. The RFP must outline clearly any requirements related to the desired format of the response. This is especially important if the RFP contains any questions for which the response is mandatory or an affirmative response is required. The RFP can stipulate that bidders not complying with such requirements will be disqualified. In most cases, it is in the interest of the local government to require that the vendor respond to bid questions directly, not by reference to marketing material or other documentation. It is difficult to evaluate responses when answers to such questions are buried in the body of the response.

Adapted from Public Technology, Inc., *The Local Government Guide to Geographic Information Systems* (Washington, D.C.: PTI, 1991), pp. 52-56.

Glossary

American Standard Code for Information Interchange (ASCII) A widely used system for encoding letters, numerals, punctuation marks and signs as binary numbers.

Archival Readable (and sometimes writeable) for a long time—from five to more than 100 years.

Back-file conversion The process of scanning and storing in digital format historical paper-based files.

Bandwidth The range of frequencies that can be passed through a channel; the capacity of a transmission medium.

Barcode A system of encoding data in machine-readable lines of variable width and configuration.

Bit A single digit in a computer binary number (1 or 0). Groups of bits make up storage units called characters or bytes.

Bit-mapped display A video monitor that stores bit patterns for digitized document images in internal memory and selectively illuminates or darkens the display in areas that correspond to light and dark pixels in the original document.

Byte A unit of computer storage holding the equivalent of a single character.

CCITT Group 3 and Group 4 International compression/decompression standards for black-and-white images specified by the Consultative Committee on International Telegraphy and Telephony (CCITT); the international standard for facsimile communications.

Compact disc A read-only optical disk available in formats for audio, data and other information. The most common type of compact disc measures 4.75 inches (120 mm) in diameter.

Compact Disc, Read-Only Memory (CD-ROM) The compact-disc format for computer data. Generally used for the storage of relatively unchanging data and/or images, such as archival files.

Compression algorithms Procedures according to which digital data sets representing images are compressed. Some of the most successful of these are contained in the CCITT Group 3 standards for digital facsimile.

Computer-Aided Acquisition and Logistics Support (CALS) A U.S. Department of Defense initiative supporting the electronic interchange of engineering documents between contractors and government agencies.

Computer Output Microfilm (COM) A system in which digital information is converted into an image on dry processed microfilm.

Computer Output to Laser Disk (COLD) A system used to control the transfer of computer-generated output to optical disk for on-line or off-line storage.

Decompression Reconstruction of a compressed image for display or printing.

De-skewing 1) The adjustment made to an image to compensate for physical distortions inherent in the system. 2) The adjustment made to an image to compensate for justification errors in scanning.

Dots per inch (d.p.i.) A measurement of resolution, e.g., the number of pixels per inch on a workstation display monitor.

Erasable optical disks A type of read/write optical disk that permits the deletion of information and the reuse of previously recorded disk areas.

Ethernet A local-area network that uses high-speed communications at up to 10 megabits per second.

FAX gateway An electronic gateway that enables incoming faxes to be converted to image data, and image data to be converted to FAX format for transmission.

FAX server A network device providing a variety of client-oriented FAX services.

Full-text search The ability to search the text content of document images. Full-text systems usually rely upon preprocessing of image files that converts them to ASCII format and then stores that ASCII data in a highly accessible form.

Geographic Information System (GIS) A system of computer hardware, software and procedures designed to support the capture, management, manipulation, analysis, modeling and display of spatially referenced data for solving complex planning and management problems.

Gigabyte One gigabyte is equivalent to one billion computer-encoded characters.

Gray scale A series of achromatic tones (tones with brightness but no hue) with varying proportions of white and black, yielding a full range of grays between white and black. In computer graphics systems with a monochromatic (black-and-white) display, variations in brightness level (gray scale) are used to enhance the contrast among various design elements.

Halftone The reprographic process in which various screens are used during printing to improve continuous-tone photograph images, creating varying sizes of black dots. Digital imaging systems employ electronic techniques to simulate that process.

Image 1) The digital representation of one side of a document. 2) To form a digital representation of a document.

Image header A packet of information that correlates an image to document information stored in the host or server and that allows for later retrieval of the image.

Indexing A method in which a series of attributes are used to uniquely define an imaged document so that it may later be identified and retrieved.

Intelligent Character Recognition (ICR) Distinguished from OCR by its ability to recognize controlled handwriting, such as handprinted letters or numbers, usually within the confines of a box or series of boxes for each individual character.

Jukebox An automatic selection and retrieval device that provides rapid, on-line access to multiple optical disks.

Kilobyte One thousand bytes. The prefix kilo- means 1,000. Also written as Kbyte.

Laser Light Amplification by Stimulated Emission of Radiation. A device for generating coherent radiation in the visible, ultraviolet and infrared portions of the electromagnetic spectrum.

Laser printer A non-impact device, common in electronic imaging systems, using laser beams to create a temporary image on a photosensitive material. This latent image is developed by applying toner particles, which are subsequently transferred and permanently fused to create the paper print.

Local-Area Network (LAN) A data communication network confined to a relatively small area, such as a building or group of buildings, and usually capable of supporting high speeds and heavy traffic volumes.

Magnetic disk A flexible or hard-disk medium used to store data in the form of minute local variations in magnetization of the disk surface.

Magnetic disk cache A directory on a magnetic disk in the Document Storage Processor (DSP) that provides storage for and quick access to frequently used documents.

Magneto-optical disk A type of erasable optical disk that uses laser optics to alter the polarization of a magnetic disk coated with a combination of iron and rare-earth transition metals.

Megabyte One million bytes. The prefix mega- means 1,000,000.

Microfiche A technique for storing multiple pages of a document on a single sheet of photographic film.

Microfilm The processed photographic film kept for later retrieval and viewing.

Micrographics A generic term that encompasses microfiche, microfilm, rolled film, aperture cards and similar technologies.

Optical Character Recognition (OCR) A device that scans printed documents and attempts to recognize the letters, numerals and other characters they contain, converting those characters to ASCII representation for computer storage.

Optical disk Platter-shaped storage medium coated with an optical recording material.

Optical storage Technologies, equipment and media that use light—specifically, light generated by lasers—to record and/or retrieve information.

Pitch The distance between tracks (track pitch) or marks (bit pitch) on an optical disk.

Pixel Abbreviation of "picture element," or one of many millions of small dots that collectively constitute the digital image. Usually referred to as number per inch (e.g., 200 pixels per inch).

Platter A large, round disk for storing information (e.g., a phonograph record).

Pre-fetching The process of building a queue of images for subsequent processing.

Resolution As applied to document scanning, the specific pattern and number of picture elements sampled by a given scanner or displayed by a graphics monitor. Resolution indicates the potential for detail in scanned images and is an important determinant of image quality.

Rotate To change the orientation of information (such as images) displayed on a workstation monitor. For example, a 90-degree rotation allows a vertical object to be viewed in a horizontal orientation.

Scanner A device that converts an original document to an electronic image.

Skewing An image condition resulting from physical distortion inherent to a monitor or from justification errors in document scanning.

Tag Image File Format (TIFF) A de facto standard file format designed to promote the interchange of digital image data; developed jointly by Aldus and Microsoft.

UNIX An operating system that has become a de facto industry standard, supported on a wide range of hardware systems from a variety of vendors. UNIX System V is the version most widely accepted in the computer industry.

Workflow management The ability to route and track electronic documents through an organization in a procedural fashion.

Write Once/Read Many (WORM) An attribute of certain optical disks. Once information is "burned" into a sector of the disk it cannot be deleted or overwritten, but can be read any number of times.

Glossary Sources

How to Speak Imaging. Blue Bell, Pa.: Unisys Corp., 1990.

National Archives and Records Administration and National Association of Government Archives and Records Administrators. *Digital Imaging and Optical Storage Systems: Guidelines for State and Local Government Agencies*. Washington, D.C.: NARA, 1991.

Wallace, Scott. *Implementing Electronic Imaging: A Management Perspective*. Warwick, Mass.: Londahl & Wallace, 1993.

Resources

Books and Reports

Association for Information and Image Management. *Legal Requirements for Electronic Imaging Systems: Audience Handbook*. Silver Spring, Md.: AIIM, 1993.

Electronic Image Management Survey: State Government: 1991 Survey Results. Wilton, Conn.: Deloitte and Touche, 1991.

Hall, George M. *Image Processing: A Management Perspective.* New York: McGraw Hill, 1991.

National Archives and Records Administration and National Association of Government Archives and Records Administrators. *Digital Imaging and Optical Storage Systems: Guidelines for State and Local Government Agencies.* Washington, D.C.: NARA, 1991.

Wallace, Scott. *Implementing Electronic Imaging: A Management Perspective.* Warwick, Mass.: Londahl & Wallace, 1993.

Selected Articles

Alsup, Mike. "Imaging Undergoes a Metamorphosis: Imaging LAN Buyer's Guide." *Network World*, 14 June 1992, pp. 41-47.

Barrett, Richard. "Client/Server Imaging Systems: A Term That's Lost Its Meaning." *Imaging World*, May 1994, p. 8.

Bowman, Teri. "Non-Proprietary Imaging: Cost-Effective Service." *Government Technology,* February 1993, p. 33.

"Building a New Image for Small Town America." *Government Imaging*, April 1994, pp. 9-10.

"Electronic Records Admissible as Evidence, Experts Say." *Government Imaging*, March-April 1994, p. 1.

"Essex County Moves to Imaging for Deeds Records." *Government Imaging,* February 1994, p. 21.

Freedman, Alan. "Imaging—A Picture Is Worth..." *Government Technology,* May 1994, p. 34.

Friedman, Rick. "Micrographics and Imaging Create the Perfect Mix." *The Office*, October 1993, pp. 38-39.

"Grumman and Empire Municipal Services to Process Parking Fines for the City of Yonkers." *Government Imaging*, October 1993, p. 12.

Hurwicz, Mike. "A Sharper Image." *LAN Magazine*, June 1994, pp. 125-132.

"Imaging Overcomes Space and Time." *Government Technology*, April 1993, p. 48.

"IRS Manages 30 Tons of Microfilm by Implementing Digital Workstations." *Imaging World*, April 1993, p. 40.

James, Fred. "Election Burden Defeated by Imaging." *Government Technology*, December 1993, p. 24.

Langner, Mark. "Don't Abandon Your Filing Cabinets Just Yet." *Network World,* 13 September 1993, pp. 71-74.

"Little Rock Scissors Paper." *Government Technology*, September 1993, p. 30.

Martin, John. "Reengineering Government." *Governing*, March 1993, p. 27.

Mechling, Jerry. "Reengineering: Part of Your Game Plan?" *Governing*, February 1994, p. 48.

Miller, Brian. "County Beats Growth with Imaging." *Government Technology*, April 1993, p. 8.

______, "Camera Captures Red-Light Runners." *Government Technology*, August 1993, p. 38.

______, "Imaged Documents and the Courts." *Government Technology*, March 1994, p. 34.

______, "Profits in Government." *Government Technology*, February 1994, p. 42.

"Milwaukee Installs Electronic System for Criminal Suspect Booking." *Government Imaging*, April 1994, p. 1.

Newcombe, Tod. "Chicago Corners Parking Violators with Imaging Net." *Network World,* 7 March 1994, p. L2.

______, "Imaging Helps Syracuse Put Police Back onto Streets." *Government Technology Executive Handbook on Imaging,* January 1992, p. 25.

______, "Workflow: The Force Within Imaging." *Government Technology*, April 1994, p. 52.

______, "The Value of Innovative Imaging." *Government Technology*, April 1993, p. 1.

"Open Systems Imaging." *Government Technology*, July 1993, p. 31.

Patton, Jim and Fritz Coburn. "County Improves Its Images." *Government Technology*, April 1994, p. 26.

"Phoenix, A Pioneer in Imaging, Wins Acclaim and Saves Millions." *Modern Office Technology*, February 1993, pp. 33-34.

Richter, M.J. "Imaging: Governing Guide." *Governing*, April 1993, pp.45-56.

______, "A Guide to Planning for Imaging." *Governing*, April 1994, pp. 71-80.

Tenalio, Michael. "Optical Disk to the Rescue." *American City and County*, March 1993, p. 10.

Wanninger, Lester A., Jr. "Systems Integration Lessons Learned in the Minnesota Imaging Project." *Inform*, November 1993, pp. 36-40.

Womack, Ralph. "Imaging Assists Investigators." *Government Technology*, August 1993, p. 30.

Periodicals

Document Imaging Report
Phillips Business Information, Inc.
7811 Montrose Road
Potomac, Maryland 20854
301/340-2100

Government Technology
9719 Lincoln Village Drive, Suite 500
Sacramento, California 95827
916/363-5000

Government Imaging
GI Communications Corp., Inc.
1734 Elton Road, Suite 200
Silver Spring, Maryland 20903-1724
301/445-4405

Governing
Congressional Quarterly, Inc.
2300 N Street, NW, Suite 760
Washington, D.C. 20037
202/862-8802

Imaging: The Imaging Industry Magazine
11 W. 21st Street
New York, New York 10010
800/999-0345

Imaging World
JW Publishing
P.O. Box 1328
Camden, Maine 04843
207/236-6267

Inform
Association for Information and Image Management
1100 Wayne Avenue, Suite 100
Silver Spring, Maryland 20910
301/587-8202

InfoTech Report
International City/County Management Association
777 North Capitol Street, NE, Suite 500
Washington, D.C. 20002

Organizations

Association for Information and Image Management
1100 Wayne Avenue, Suite 100
Silver Spring, Maryland 20910
301/587-8202

International City/County Management Association
777 North Capitol Street, NE, Suite 500
Washington, D.C. 20002
202/289-4262

National Archives and Records Administration
Office of Policy and Information Resources Management Services
8601 Adelphi Road
Room 3200
College Park, Maryland 20740-6001
301/713-6730, ext. 263 (Barry Roginski)

National Association of Counties
440 First Street, NW
Washington, D.C. 20001
202/393-6226

National League of Cities
1301 Pennsylvania Avenue, NW
Washington, D.C. 20004
202/626-3000

Public Technology, Inc.
1301 Pennsylvania Avenue, NW
Washington, D.C. 20004
202/626-2400